Mapping Middle-earth

Perspectives on Fantasy

Series Editors
Brian Attebery (Idaho State University, USA)
Dimitra Fimi (University of Glasgow, UK)
Matthew Sangster (University of Glasgow, UK)

The first academic series with an exclusive critical focus on Fantasy, *Perspectives on Fantasy* publishes cutting-edge research on literature and culture that brings sophisticated discussion to a broad community of debate, including scholars, students, and non-specialists.

Inspired by Fantasy's deep cultural roots, powerful aesthetic potential, and reach across a broad range of media – from literature, film and television to art, animation and gaming – *Perspectives on Fantasy* provides a forum for theorising and historicising Fantasy via rigorous and original critical and theoretical approaches. Works in the series will cover major creators, significant works, key modes and forms, histories and traditions, the genre's particular affordances, and the ways in which Fantasy's resources have been drawn on, expanded and reconfigured by authors, readers, viewers, directors, designers, players, and artists. With a deliberately broad scope, the series aims to publish dynamic studies that embrace Fantasy as a global, diverse, and inclusive phenomenon while also addressing oversights and exclusions. Along with canonical Anglophone authors and texts, the series will provide a space to address Fantasy creators and works rooted in African, Asian, South American, Middle Eastern, and indigenous cultures, as well as translations and transnational mediations.

The series will be alive to Fantasy's flourishing fan cultures, studying how audiences engage critically and affectively and considering the ease with which participants in Fantasy communities move from being readers and watchers to players, writers, and artists.

Editorial Board Members
Catherine Butler (Cardiff University, UK)
Paweł Frelik (University of Warsaw, Poland)
Rachel Haywood Ferreira (Iowa State University, USA)
Robert Maslen (University of Glasgow, UK)
Ebony Elizabeth Thomas (University of Pennsylvania, USA)
Anna Vaninskaya (University of Edinburgh, UK)
Rhys Williams (University of Glasgow, UK)
Helen Young (Deakin University, Australia)

Titles in this Series
Queering Faith in Fantasy Literature: Fantastic Incarnations and the Deconstruction of Theology, Taylor Driggers
William Hope Hodgson and the Rise of the Weird: Possibilities of the Dark, Timothy S. Murphy
Imagining the Celtic Past in Modern Fantasy edited by Dimitra Fimi and Alistair J. P. Sims

Forthcoming Titles
Justice and the Power of Wonder in 21st-Century Fairy Tales, Cristina Bacchilega and Pauline Greenhill
Reading Tolkien in Chinese: Religion, Fantasy and Translation, Eric Reinders
Fantasy and the Politics of Subversion: Speculative Writing in Colonial India, Mayurika Chakravorty
Femslash Fanfiction: Analysing Queer Time in Swan Queen Fan Narratives, Alice Kelly

Mapping Middle-earth

*Environmental and Political Narratives in
J. R. R. Tolkien's Cartographies*

Anahit Behrooz

BLOOMSBURY ACADEMIC
LONDON • NEW YORK • OXFORD • NEW DELHI • SYDNEY

BLOOMSBURY ACADEMIC
Bloomsbury Publishing Plc
50 Bedford Square, London, WC1B 3DP, UK
1385 Broadway, New York, NY 10018, USA
29 Earlsfort Terrace, Dublin 2, Ireland

BLOOMSBURY, BLOOMSBURY ACADEMIC and the Diana logo are trademarks of Bloomsbury Publishing Plc

First published in Great Britain 2024

Copyright © Anahit Behrooz, 2024

Anahit Behrooz has asserted her right under the Copyright, Designs and Patents Act, 1988, to be identified as Author of this work.

For legal purposes the Acknowledgements on p. ix constitute an extension of this copyright page.

Series design by Rebecca Heselton
Cover illustration by Rebecca Heselton

All rights reserved. No part of this publication may be reproduced or transmitted in any form or by any means, electronic or mechanical, including photocopying, recording, or any information storage or retrieval system, without prior permission in writing from the publishers.

Bloomsbury Publishing Plc does not have any control over, or responsibility for, any third-party websites referred to or in this book. All internet addresses given in this book were correct at the time of going to press. The author and publisher regret any inconvenience caused if addresses have changed or sites have ceased to exist, but can accept no responsibility for any such changes.

A catalogue record for this book is available from the British Library.

Library of Congress Cataloging-in-Publication Data
Names: Behrooz, Anahit, author.
Title: Mapping Middle-earth : environmental and political narratives in J.R.R. Tolkien's cartographies / Anahit Behrooz.
Description: London ; New York : Bloomsbury Academic, 2024. | Series: Perspectives on fantasy | Includes bibliographical references and index.
Identifiers: LCCN 2023030907 (print) | LCCN 2023030908 (ebook) | ISBN 9781350290761 (hardback) | ISBN 9781350290808 (paperback) | ISBN 9781350290778 (adobe pdf) | ISBN 9781350290785 (epub)
Subjects: LCSH: Tolkien, J. R. R. (John Ronald Reuel), 1892–1973–Criticism and interpretation. | Cartography in literature. | Maps in literature. | LCGFT: Literary criticism.
Classification: LCC PR6039.O32 Z56624 2024 (print) | LCC PR6039.O32 (ebook) | DDC 823/.912–dc23/eng/20230726
LC record available at https://lccn.loc.gov/2023030907
LC ebook record available at https://lccn.loc.gov/2023030908

ISBN:	HB:	978-1-3502-9076-1
	PB:	978-1-3502-9080-8
	ePDF:	978-1-3502-9077-8
	eBook:	978-1-3502-9078-5

Series: Perspectives on Fantasy

Typeset by Integra Software Services, Pvt. Ltd.
Printed and bound in Great Britain

To find out more about our authors and books visit www.bloomsbury.com and sign up for our newsletters.

Contents

Acknowledgements	ix
Introduction	1
Space, Power and Critical Cartography	5
Literary Maps	12
Structure and Overview	18
1 Political Map-making	21
Medieval Cartographic Practices	26
Modern Cartographic Practices	35
Tolkien's Cartography	44
Map I: I Vene Kemen	44
Map II: The 'Ambarkanta' Diagrams and Maps	46
Map III: Thror's Map	49
Map IV: The Middle-earth Map	53
Map V: Map of Rohan, Gondor and Mordor	59
2 Environment	63
Navigating the Human, Non-human and Posthuman	65
Tom Bombadil and the Non-human	68
Mapping the Human and Non-human in Middle-earth	72
Stewardship	78
Environmental Destruction	84
Non-human Agency	88
3 Geology and Time	97
Deep Time	98
Middle-earth's Geology	103
Mapping Geology and Geologizing Maps	112
Fixing Experiences of Time	118
Mapping Anthropological Change	126

4 Imperialism and Race	133
The Politics of Land and Map	136
(Dis)possessing Middle-earth's Lands	141
The Threshold Space	143
Mutual Vulnerability and Racialization	151
Narratives of Imperialism	160
Conclusion	171
Bibliography	176
Index	184

Acknowledgements

It is appropriate, in many ways, that this study into cycles of crisis, extraction and exploitation was written in this current moment, when academia itself has increasingly become defined by the same. The research for this book was undertaken firstly as an unfunded PhD project at the University of Edinburgh and subsequently, without institutional affiliation, around a full-time job that financed its completion. I am stating this here both to highlight – in a book so interested in the materiality of power – the material conditions that informed the production of this and intellectual labour more broadly (and that are so often invisibilized within the contemporary academy), and to acknowledge the support I received both within and also beyond the remits-as-stands of the university institution.

First and foremost, I owe everything good in this book to my wonderful, encouraging and extraordinarily knowledgeable PhD supervisor Dr Anna Vaninskaya. Many thanks also to my second supervisor Dr Sarah Dunnigan, and to Dr Jonathan Wild, who read early sections of the second chapter and who encouraged me to 'run with maps'.

I am incredibly grateful to my viva committee Professor Adam Roberts and Professor David Farrier for their kindness and advice – especial thanks to David for suggesting the reading that made the second chapter what it is.

Immense thanks are due to Dr Dimitra Fimi for much patience and mentorship over the years, Dr Judy Kendall for first encouraging me to send this proposal to Bloomsbury, and the readers and peer reviewers who engaged with the original proposal and early drafts.

All my love and thanks to Dr Marian Chan, to whom I owe the years of my PhD and much beyond it. Thank you also to my editor Rosamund West and my colleagues at *The Skinny* for providing time and space to write.

Thank you forever to Bridget and Maria for reading and listening, and Heather for writing the emails.

Finally, thank you to my family, and most especially my parents Ali and Gita Behrooz, who believed in this more than anyone.

Introduction

Twelve years after its initial publication, Scottish author Robert Louis Stevenson recalled how he came to write *Treasure Island* (1882), the novel that had propelled him to fame and (not quite) fortune. As with his fictional hero Jim Hawkins, who discovers a treasure map in a pirate's chest and is thrust onto a path of adventure and intrigue, for Stevenson too it began with a map. Convalescing at a cottage in Aberdeenshire after a short illness, Stevenson recollects playing with his landlady's artistically minded young son, creating a small picture gallery out of paper and ink and watercolours. Largely taking on the role of museum guide, Stevenson would on occasion contribute his own creative endeavours to the collection: it was on one of these occasions that he drew an 'elaborately and ... beautifully colored' map that he subsequently labelled 'Treasure Island' (1895, 5). Stevenson describes how, while gazing at the map, suddenly 'the future characters of the book began to appear there visibly among imaginary woods; and their brown faces and bright weapons peeped out upon me from unexpected quarters, as they passed to and fro, fighting, and hunting treasure, on these few square inches of a flat projection', adventures that Stevenson hastily put to paper (1895, 5–6). Even after *Treasure Island* emerged as a complete novel, the map remained Stevenson's focal point for both the story and its history; attempting to explain the crucial connection between the map and the narrative, Stevenson advises every author 'in the beginning to provide a map', explaining 'I have said the map was the most of the plot. I might almost say it was the whole ... The tale has a root there; it grows in that soil; it has a spine of its own being ...' (1895, 10–11).

For the reader of J. R. R. Tolkien, the story also frequently starts, and very often even ends, with a map. His novels *The Hobbit* (1937) and *The Lord of the Rings* (1954–55) are bookended by cartographic prints that depict different parts of the geography and topography of Middle-earth, drawn by himself and his son Christopher Tolkien. Tolkien's 'Silmarillion' writings, unpublished in his lifetime and subsequently edited and published by Christopher as the edited *The Silmarillion* (1977), the twelve volumes of *The History of Middle-earth* (1983–96) and the collected *Unfinished Tales* (1980), similarly feature sketch maps and elaborate charts depicting the various lands of Middle-earth and the broader sub-created world of Arda in previous, mythological Ages. These maps have primarily been presented, read and understood as paratextual devices; as Ricardo Padrón argues:

> [t]hey pull us down to earth (to Middle-earth, that is), inviting us to consider the landscape from the perspective of someone traveling through it. We follow the road through the forests across the mountains, along the rivers, sometimes tracing the paths of Frodo and the others, and sometimes forging our own way …
>
> (Padrón 2007, 274)

Here, Padrón highlights the ways in which Tolkien's maps have long facilitated the reader's visualization of both the fictional geography and the narrative trajectory of his legendarium, providing an additional layer of world-building verisimilitude that allows for a deeper engagement with the shifting lands of Middle-earth. Yet these maps also have a further, significant function that has been largely critically neglected. Much as Stevenson considered the map the whole of the plot, and the catalyst that allowed his narrative to germinate, Tolkien's maps also play a central role within the narrative of the texts themselves, working as examples of a broader fictional cartographic practice that contributes to the social, cultural and political character of his sub-created world. Crucially, Tolkien's maps extend beyond objects intended merely to ease the external reader's understanding of the narrative; rather, they indicate a sub-created tradition of cartography that articulates particular relationships between the map-maker, the map-reader and what is being mapped, which are expressed both through the physical maps and in the wider legendarium.

Tolkien was, of course, conscious of his maps as an illustrative complement to the text: in a 1954 letter to author and friend Naomi Mitchison, shortly before the publication of *The Lord of the Rings*, Tolkien mirrors Stevenson's sentiments, explaining, 'I wisely started with a map, and made the story fit (generally with meticulous care for distances). The other way about lands one in confusions and impossibilities …' (2006, 177). Tolkien is referring here to the paratextual function of the map and his desire to have it accurately represent the events of the novel; yet the reciprocal relationship that he highlights between map and story is integral to understanding how his cartography also tessellates with broader narratives in the legendarium that centre on the connection between humans and the wider world. In as much as the story is made to fit the map, the map also fits the story, enabling it to speak to these wider thematic concerns and the relationships between the mapper and the mapped.

Specifically, in this book, I argue that these relationships frequently embody power dynamics that are inherent between people and their surroundings – focusing on environmental damage, temporal and spatial control of land, political conflict and imperialist violence – and that are notably inextricable from the act of cartography itself. It is therefore essential to understand these maps within a broader tradition of political and politicized cartography, as has been theorized in the field of critical cartography by theorists such as J. B. Harley, Denis Wood and John Fels, who position maps as socially constructed texts that can be employed and deployed as a tool of power. Although they are stylistically, conceptually and functionally varied objects, what unites Tolkien's maps is their adjacency to these exercises of power within the narratives, both as expressions and actants of these exercises. Tolkien's fictional cartography thus has a key hermeneutic role in investigating these narratives of power, yet significantly, it also acts as a self-reflexive consideration of the inherent politics of mapping. This is the central concern of this book: I intend to demonstrate how Tolkien's maps are both representative and productive of exercises of power and how they demonstrate a pervasive and inescapable pattern of domination that is intrinsic to cartography and – more widely – human relationships with their surroundings.

This study therefore intends to move on from the critical characterization of Tolkien's Middle-earth maps as illustrative or paratextual devices, and instead examine them as material examples of a broader tradition of fictional cartography that is embedded within his sub-creation, and that intersects with numerous political issues in the text. It is essential to understand the ways in which Tolkien positions cartography as an inherently political act that embodies a desire for totalizing understanding and control of its subject matter; this problematizing of external control then enables a critique of other harmful contemporary engagements with land that manifest in, but also move beyond, cartography, specifically the environmental damage caused by industrialization; the tension between human and non-human temporalities catalysed by the discovery of deep time; and the ecological and bodily costs of political violence and imperialism, brought into extreme relief by the activities of the British Empire. The intention is to place Tolkien within an explicitly contemporary and theoretical context from which he has hitherto been largely excluded;

each chapter therefore engages with ecocritical and postcolonial frameworks (as will be discussed in greater detail below), as well as with geographical and cartographic theory that enables a rigorous consideration of the politics of space within Tolkien's sub-creation. As each chapter is thematically distinct, theoretical frameworks will largely be set up at the start of each individual chapter; instead, this Introduction intends to unpack the broader conceptualization of literary cartography, and to demonstrate the ways in which positioning Tolkien's maps as diegetic devices that ought to be read through an explicitly cartographic lens has critical significance.

This diegetic reading of Tolkien's maps is not to be assumed, given how they are frequently read through a paratextual lens by Padrón and others, discussed below. Of the full cartographic corpus, only one is an explicitly diegetic map: Thror's Map from *The Hobbit*, which the dwarves use to locate and enter the Lonely Mountain, is both a textual object within the narrative itself, read and interpreted by Thorin's Company, and an external reader-oriented device, originally included as *The Hobbit*'s endpaper. However, there is definitive justification to consider all of Tolkien's Middle-earth maps as products of an internal cartography, which consequently frames them as tools of power for the cultures that produce them within the text. In his seminal study on fantasy cartography, Stefan Ekman points out that the placement of 'A Part of the Shire' in between the Prologue and the first chapter of *The Lord of the Rings*, and thus in the middle of the narrative, means that 'rather than providing a paratextual threshold, it is evidently part of the narrative document' (Ekman 2013, 38).[1] In the context of Tolkien's extended conceit that *The Lord of the Rings* and *The Hobbit* are edited texts sourced from Bilbo's Red Book of Westmarch, Ekman posits that Tolkien frames the map as another artefact from the Red Book of Westmarch, making it a diegetic map which would have been created in Middle-earth and used by the characters. This argument can then be extended to the other maps that appear within the legendarium. As Tolkien consistently frames these texts as 'found' narratives that he has simply translated and edited,

---

[1] Prologues themselves are of course frequently considered paratextual devices; Gerard Genette groups prologues along with other preambulatory remarks such as introductions and avant-propos under the term 'preface', and argues that they are all types of paratext (Genette 1997, 161). Ekman does not explore this distinction; however, Tolkien's Prologue clearly situates itself within the wider conceit of *The Hobbit* and *The Lord of the Rings* as an edited version of Bilbo's Red Book of Westmarch, explaining in the first paragraph: 'Further information will also be found in the selection from the Red Book of Westmarch that has already been published' (2008a, 1). Although the prologue is physically separated from the main narrative of *The Lord of the Rings*, its engagement with this conceit means it participates in the narrative more so than a typical prologue. It also follows that material that appears after it, such as 'A Part of the Shire', also engages with this conceit and – as Ekman argues – is not excluded entirely from the narrative.

the other maps in his corpus can effectively be read as reproductions of Middle-earth's cartography found in the Red Book of Westmarch.

Starting from the assumption that these maps can be understood diegetically, this Introduction will first define the mechanisms of power at play in political and politicized cartography through an examination of Michel Foucault's conceptualization of power, and the ways in which it has been adopted within the theory of critical cartography. I will then give an overview of the various critical considerations of literary cartography, focusing on those who are interested in authorially produced material maps, including Padrón, Ekman and Mark J. Wolf; and geographers such as David Cooper, Keith D. Lilley and Andrew Thacker, who advocate for an explicitly cartographic framework in literary studies that engages with the political nuances of mapping, which forms the basis of my own approach. In doing so, I intend to demonstrate the necessity of engaging with frameworks such as critical cartography that emphasize the ways in which Tolkien's texts can be read as politically rich and critical modes of writing, and as products of his contemporary culture's anxieties that remain relevant to our present theoretical concerns. In placing Tolkien so directly in conversation with discourses of power – from the field of critical cartography to the narratives of power over land that it articulates, including the emerging and still ongoing ecological crisis, anxieties around non-human deep time and the ravages of colonialism and imperialism – this book intends not merely to draw attention to the radical nature of Tolkien's writing but to insist on a radical reading of it.

Space, Power and Critical Cartography

This book's engagement with cartography is based on a primarily Foucauldian conceptualization of power, in which power is defined as 'forms of domination [and] … subjection … which have their own ways of functioning, their own procedure and technique' (2007, 156). For Foucault, these forms of domination do not necessarily manifest as legislation or prohibition; instead, he argues, Western societies have largely imagined and deployed power in a negative, restricted way. Instead, Foucault claims, power needs to be understood heterogeneously: rather than an abstract concept of 'power', it is necessary to conceptualize 'powers' that are produced within specific historical and geographical contexts, that are in turn 'producers of an efficiency, an aptitude, producers of a product' (2007, 157). Tally categorizes Foucault's theory of power as 'pervasive, capillary, and productive' (2013, 123), which draws appropriate attention to the ways in which Foucault perceived these exercises of power as deeply entrenched within and inextricable from social and cultural practices.

This totalizing presence of power is made evident in Foucault's examination of power/knowledge relations in historiography and sexuality. Mark Poster argues that Foucault sees history as a 'means of controlling and domesticating the past in the form of knowing it' (1982, 119): this is discussed in *Archaeology of Knowledge* (originally published in France under the title *L'Archeologie du savoir* by Editions Gallimard 1969), where Foucault critiques the historian's inability to view the past as discontinuous to the present. Instead, Foucault says, the historian explains and recreates the past in relation to the present; the authority produced through this discourse gives the historian power not only over how the past is represented but also over how the present is configured. As Foucault explains:

> Continuous history is the indispensable correlative of the founding function of the subject: the guarantee that everything that has eluded him may be restored to him; the certainty that time will disperse nothing without restoring it in a reconstituted unity; the promise that one day the subject – in the form of historical consciousness – will once again be able to appropriate, to bring back under his sway, all those things that are kept at a distance by difference, and find in them what might be called his abode …
>
> (2002, 13)

This approach centres the historian within the production of history, creating a historical epistemology that is inextricably bound up in the power of its producer. This relationship between the production of knowledge and the production of power is also discussed in *The History of Sexuality* (originally published in France under the title *La volonté de savoir* by Editions Gallimard 1976), specifically through the act of confession. Foucault argues that 'Western societies have established the confession as one of the main rituals we rely on for the production of truth' (1978, 58), with sexual preference and activity becoming a particularly 'privileged theme of confession' (1978, 61). Beginning with the confessional of the Catholic Church, Foucault illustrates how the impetus to confess the sexual self has permeated beyond the religious, manifesting in relationships between parents and children, medical professionals and patients, and teachers and students. These acts of confession encapsulate the ways in which knowledge is appropriated as a tool for power: the confessor reveals the truth about their sexuality and in doing so communicates knowledge, which is then categorized, administered and controlled by the listener, creating a framework of power hierarchy. Both examples – Foucault's study of historiography and the confessional – encapsulate his insistence on power as heterogeneous: each is dependent on particular historical and social contexts, yet both demonstrate a pervasive relationship and deeply knotted relationship between knowledge and power.

Crucially, Foucault also examines this relationship in matters of space. Much of Foucault's work was concerned with the ways in which power is deployed spatially, to the extent that Gilles Deleuze, in his review of *Discipline and Punish* (1977), christened Foucault the 'new cartographer' of social spaces (qtd in Tally 2013, 120). Tally argues that Foucault's earliest work *Madness and Civilization* (1964) demonstrates how the birth of the modern mental asylum was 'part of a powerful and nuanced centralization, classification, and organization of space' (2013, 124), in which individuals were categorized, medicalized and placed in suitable spaces. In *Discipline and Punish*, meanwhile, Foucault does indeed map how power can spatially be configured for punitive means through his discussion of the Panopticon. The Panopticon model was first theorized by Jeremy Bentham, who argued that a prison that would allow all inmates to be watched by a single watchman – without knowing if they were being observed in that moment or not – would lead to a system of total control, as inmates would be effectively forced to constantly 'watch' themselves and regulate their own behaviour. Foucault builds on Bentham in order to draw attention to the power dynamics of the very act of watching, that is to say, of accumulating knowledge, that is enabled by the architecture of the Panopticon. Specifically, Foucault emphasizes how power can be manifested simply through the illusion of constant surveillance, and how power can be removed through the illusion of constant visibility:

> Hence the major effect of the Panopticon: to induce in the inmate a state of conscious and permanent visibility that assures the automatic functioning of power. So to arrange things that the surveillance is permanent in its effects, even if it is discontinuous in its action; that the perfection of power should tend to render its actual exercise unnecessary ...
>
> (*Discipline and Punish* 205)

As Thomas Flynn argues, the Panopticon is a spatialized account of a broader critique of the '"disciplinary" society of the modern age', which encompasses 'the omnipresence of surveillance devices, the vulnerability of our various communications systems to external review and interpretation, and the insinuation of authorities into the most private portions of our personal lives …' (2016, 60). By expressing these concerns through the architectural model of the self-regulating prison, Foucault demonstrates the complicity of space within these broader exercises of power, and how these spaces too can be used to dominate and subjugate. Whether through historiography and the creation of other sociocultural narratives, or through the formation and categorization of space, Foucault demonstrates, as Tally termed it, the pervasive and capillary nature of power, and its appropriation and exploitation of forms of knowledge for its own growth.

Although Foucault does not directly address cartography in these considerations of how power is practised, his conceptualization of power as a force of domination that does not prohibit but rather produces and that is inextricably linked with epistemological practices speaks distinctly to similar cartographic traditions. In particular, Foucault's framework has been employed by critical cartographers, a field of cartographic criticism that is rooted in critical theory and that seeks to illuminate the ways in which maps encode practices of power. The historic methods by which maps deploy these codes of power is the subject of chapter one and so will not be examined in full here; however, it is worth highlighting the key tenets of critical cartography here, and the ways in which they theorize the broadly and inherently power-implicit practices of the cartographic act. Harley, one of the leading critics in this field, draws directly on Foucault's theorization of the relationship between power and knowledge, arguing that just as the historian configures the past through the lens of the present, so too does the 'surveyor, whether consciously or otherwise, replicat[e] not just the "environment" in some abstract sense but equally the territorial imperatives of a particular political system' (1988, 279), so that even the most seemingly objective maps are bound up in the systems of power in which they are created and deployed. This inescapable shaping by power renders all maps 'socially constructed form[s] of knowledge' that produce and reproduce political discourse through both their content and circuit of communication (1988, 277). Harley does not define politics, but his use of it effectively encompasses large-scale exercises of power. For Harley, it is imperative that cartographic theory take a deconstructionist approach to the study of the map in order to break the assumed link between reality and representation that the map presents, which will in turn reveal the 'invisible or implied' systems of power at work in the map (2001, 152).

In 'Maps, Knowledge and Power' (1988), Harley approaches this through a parallel consideration of the map's political context and the symbolic potential of its iconography to reflect this context, rather than represent an abstract truth. In an analysis of maps and property rights in early modern Europe, for example, Harley argues that local maps are a product of certain 'long-term structural changes of the transition from feudalism to capitalism' (1988, 285), with the new economic system and its geographical division of labour being enabled by the map's representation of said geography. Accurate, large-scale maps permitted a more thorough and codified exploitation of the land, its agricultural potential and its tenants. Through the division of the land into allotments, the charting and reclamation of previously wild hills and moors, and the delineation of the land in precise measurements and scales, Harley argues that 'the surveyor ever more frequently walks at the side of the landlord in spreading capitalist forms of agriculture' (1988, 285). Much as the clock brought a regimented structure to workers' experience of time, so too did the map impose what Harley terms 'space discipline' upon the land and its inhabitants (1988, 285). For Harley, the centrality of the map to implementing land rights in the shift to capitalism is emblematic of how cartography not only represents broader political movements and exercises of power, but how it works symbiotically alongside them to enable these dynamics.

Harley expands on this argument in 'Deconstructing the Map' (2001), further drawing attention to the ways in which maps are constructed by their political contexts, and the importance of deconstructing and laying bare these enmeshments of power. Harley argues that maps have both external and internal power: externally, power is exerted on cartography through the demands of patrons, and through the ways in which the maps are employed. However, the power internal to cartography aligns the map more distinctly with Foucault's conceptualization of power not as judicial but as explicitly productive. The map's very methodology of representation and the ways in which elements of the landscape are included, excluded, categorized and simplified produce a knowledge of the land which in turn creates power, power that is 'not generally exercised over individuals but over the knowledge of the world made available to people in general' (Harley 2001, 112). The cartographer aligns exactly with Foucault's figure of the historian, in that the power created by these forms of knowledge is not deployed directly upon an individual, but rather through the communication of the knowledge itself, which becomes enmeshed and normalized within the social structure. The knowledge that the map articulates comes to inform totalizing understandings of geography, sociology and political relations, which in turn enables particular power dynamics to be formed and exercised.

Wood builds on Harley's influential approach by demonstrating the inherent subjectivity of every map, that is to say, the biases it encodes based on its author and its intended audience. Wood defines power as 'the ability to do work', arguing that maps work by serving these specific interests, thereby functioning as a continual exercise of power (1992, 1). Similar to Harley, Wood takes a Foucauldian approach by exposing the impossibility of an objective map – much as Foucault confronted the supposed objectivity of created history – by arguing that the knowledge which produces the map is always, invariably, socially and personally constructed. Wood claims that:

> *knowledge of the map is knowledge of the world from which it emerges* – as a casting from its mold, as a shoe from its last – isomorphic counter-image to *everything* in society that conspires to produce it. This, of course, would be to site the source of the map in a realm more diffuse than cartography; it would be to insist on a *sociology* of the map …
>
> (1992, 18, emphasis in original)

Wood demonstrates the undeniable relationship between the map and the sociopolitical context it is produced in by highlighting the first satellite map of the Earth created by Tom Van Sant, a cartographic model that should theoretically, according not only to the scientific means by which it was produced but also the ideology of scientific objectivity which it insists on and represents, be free of the biases or interests of its maker. Drawing on Roland Barthes' claim that a photograph be free of a 'code intervening between the object and its message' (qtd in Wood 1992, 51), Wood demonstrates how Van Sant's photographic map of the Earth fails this requirement: the map is composed of numerous fragmented satellite shots of the Earth that were stitched together; it is hand-tinted in places in order to accentuate the colour codification expected of a world map, where green represents land and blue represents water; and, as with the infamous Mercator projection, the surface of the Earth has been manipulated in order to facilitate the change from sphere to flat surface. Each of these amendments demonstrates the deliberate activity of the map-maker and the ways in which they mould the map to fit their interests, thereby creating a code that intervenes between the reality of the object (in this case, the Earth) and the message (the map). Similarly, in a collaborative study with Fels, Wood examines cartographic renditions of nature, which Fels and Wood argue is a supposedly neutral, that is to say non-human, territory, in order to demonstrate the inescapable presence of the map-maker and the map as 'nothing more than a vehicle for the creation and conveying of authority about, and ultimately over, territory' (2008, 7). The continued relevance of Harley, Wood and Fels in the field of geographical and cartographical criticism has been commented on recently by Kitchin, Chris Perkins and Martin Dodge, who argue that their 'avowedly political' approach is integral for moving beyond the conceptualization of maps as purely representational objects and instead conceiving what they term a 'post-representational cartography', which understands the ways in which maps constitute political relations in matters as diverse as colonialism, national identity, bureaucracy and gender (2009, 10).

That these are issues that intersect not only with social and political theory but also with literary studies has been commented on by certain critics, who have noted the literary potential of critical cartography as a methodological tool. Cooper argues that research interested in bridging the gap between geocriticism – criticism engaged with issues of spatiality, understood broadly – and the humanities needs to be 'predicated upon a self-reflexive engagement with geographical thinking and practices rather than an uncritically imprecise reliance on spatial vocabularies and discourses' (2012, 30). Specifically, Cooper argues that scholarship engaged with any form of literary cartography needs to be informed by the work of key cartographic critics such as Harley and Wood; work which comprehends the political as well as the representational implications of cartography, even when fictionalized. Cooper's insistence on a critical cartographic framework is motivated by what Tania Rossetto terms a 'decartographization' of the literary field, in which maps and mapping become largely regarded as metaphorical practices, thereby stripping them of the power relations that Harley and Wood argue are integral to mapping (2014, 517). Stephen Daniels et al. also argue that mapping 'as a term of cultural description in the arts and humanities has moved beyond the practice of cartography to a broader, metaphorical sense of interpreting and creating images and texts' (Daniels et al. 2011, xxx), while Melba Cuddy-Keane insists that 'we need constantly to examine the literal ground on which these metaphors depend' (2002, 8).

Regarding this relationship between maps and literature, there is a marked movement toward a return to maps as material artefacts as essential for further unpacking politics of space in literary narratives; mapping becoming a concrete political praxis rather than metaphor or a methodology. Lilley argues that '[a]t a time when figurative and metaphorical "mappings" are becoming particularly prominent in other humanities areas, such as literary criticism and philosophy, it is perhaps worth underlining the benefits of still thinking about maps and "map-making" in a more conventional and literal sense' (2011, 31). This was also a directive raised by Thacker, who suggested that '[r]ather than only treating "mapping" as a metaphor it seems important to return to the map as a set of material signs, and to understand what is at issue when a text employs an actual map as a component of the narrative' (2005, 64). Thacker argues that what he terms a critical literary cartography must return to the maps presented in texts, in order to analyse cartography as an example of visual culture within narrative. What follows in this book is positioned precisely in the methodological framework put forward by these critics, not only by considering Tolkien's maps through a critical cartographic lens, but insisting on the hermeneutic value in aligning literary studies with an explicitly geographical, and non-metaphorical, engagement with cartography. This interdisciplinary approach is vital, I argue, in not only outlining the political complexity of Tolkien's texts as informed by the accompanying maps but also in further demonstrating the constructed nature of maps even within fictional works, revealing the power relations they encode and enable. Tolkien's Middle-earth map corpus is a text that is doubly produced: both produced by its author and produced by actants within the fictional text, so that it comes to embody a complex and deeply contemporary system of knowledge production, social construction and political value. It is through reading the legendarium's maps through this lens that we can begin to understand both Tolkien's incisive grasp of methods of knowledge and power production, and also the specific narratives of power that thread through the legendarium and intersect with its cartography. The legendarium, I argue, is a deeply political text: it is only in treating it as such, and opening up the potential for such readings, that we can come to understand the extent of consideration and critique of power.

Literary Maps

Employing critical cartographic frameworks in the analysis of literature is particularly necessary given the abundance of literary maps available that span across authors, periods and genres. Diana Wynne Jones' satiric overview of the fantasy genre, *A Tough Guide to Fantasyland* begins, 'Find the MAP. It will be there. No Tour of Fantasyland is complete without one' (2004, 1). Written in 1997, Jones' tongue-in-cheek claim reflects how the fantasy map had at this point become a generic cliché, appearing in texts as diverse as Ursula Le Guin's *Earthsea Quartet* (1964–1990), Terry Pratchett's *Discworld* novels and Norton Juster's *The Phantom Tollbooth* (1961). Its fixed presence within the genre is frequently credited to Tolkien's prolific and iconic cartographic production: Ekman argues that maps have become 'almost obligatory' due to the popularity of Tolkien's novels; Farah Mendlesohn claims that 'Tolkien set the trend for maps ...' (2008, 14); and R. C. Walker explains that not only did Tolkien's maps 'set a high standard, they seem to have created an interest in fantasy maps ... so that a map has almost become de rigeur in new and reprinted fantasy' (1981, 37). While Tolkien was certainly key to the genre's twentieth-century manifestation, however, literary maps – that is to say, maps that were printed alongside the text they illustrated – had existed for centuries previously.

Disregarding biblical maps (which will be discussed in detail in Chapter 1), the first map illustrating a fictional place from a creative narrative is Sandro Botticelli's map of hell, which forms part of the 1485 illustrated manuscript of the *Divine Comedy*.[2] The map depicts hell as a series of cascading rings, each characterized according to Dante's conceptualization of the nine circles of hell. Meanwhile, the first fictional map to be designed specifically for the narrative and be printed alongside its text upon publication appeared some thirty years later in Thomas More's *Utopia* (1516). John Bunyan published *The Pilgrim's Progress* in 1678; although the original text did not contain a map, Christian's journey through Slough of Despond, the Valley of the Shadow of Death and towards the Celestial City has been mapped countless times since. A century later, Daniel Defoe's sequel to his popular *Robinson Crusoe* (1719), *Serious Reflections During the Life and Surprising Adventures of Robinson Crusoe: With his Vision of the Angelick World* (1720), featured a map of the desert island. A few years later, Jonathan Swift's satire *Gulliver's Travels* (1726) was similarly printed alongside a series of maps depicting the fictional lands of Lilliput and Brobdingnag, as well as Japan. These maps were striking for abandoning the more illustrative tendencies of *Utopia* and *Robinson Crusoe*, and tending towards a more simplified topographic style.

The growing interest in travel and adventure literature propelled the creation of several more famous literary maps in the nineteenth century: Johann David Wyss' *Der Schweizerische Robinson* (1812), first translated into English as *The Swiss Family Robinson* in 1814, features a map of 'New Switzerland', the island upon which the family is marooned; Jules Verne's subsequent version of the shipwreck narrative, *L'Île Mystérieuse* (1874) – published as *The Mysterious Island* in 1875 – includes a topographic map of Île Lincoln; Lewis Carroll's nonsense

---

[2] I want to credit Huw Lewis-Jones' *The Writer's Map* for documenting many of the maps in this section.

poem *The Hunting of the Snark* was published in 1876 alongside a blank map;[3] the writing of Stevenson's *Treasure Island*, as briefly discussed above, was rooted in the conception of the treasure map; H. Rider Haggard's colonial adventure story *King Solomon's Mines* (1885) included a roughly drawn map presented as a found document; Arthur Morrison's *A Child of Jago* (1896) features a sketch map of the slums of East London; while William Morris' *The Sundering Flood* (1897), one of the first works of modern fantasy, features a frontispiece map of the area surrounding the eponymous river.

By the twentieth century, literary cartography had become relatively prevalent in comparison to its latent beginnings: J. M. Barrie's *The Little White Bird* (1902), his precursor to *Peter Pan* (1904), featured 'The Child's Map of Kensington Gardens'; in the same year, Rudyard Kipling's *Just-So Stories* featured a detailed, illustrative map of the Amazon River; in 1912, Arthur Conan Doyle's *The Lost World* featured a hand-drawn chart of the Maple-White Land complete with annotations including 'here we saw great elk' and 'Central Lake (sandbanks and monsters)'; also in 1912, Thomas Hardy published a collected edition of his works complete with a map of Wessex; in America, the eighth book in Frank L. Baum's Wizard of Oz series, *Tik-Tok of Oz* (1914) featured two maps, one of Oz and one of the broader continent where Oz and neighbouring magical lands were located; E. H. Shepard famously provided illustrations for A. A. Milne's *Winnie-the-Pooh* (1926), along with a map of Hundred Acre Wood (possibly due to the popularity of his *Winnie-the-Pooh* illustrations, Shepherd also drew a map for the 1931 edition of Kenneth Grahame's *The Wind in the Willows* (1908)); the first edition of Arthur Ransome's *Swallows and Amazons* (1930) features a map of the Lake District on its dust jacket; William Faulkner's *Absalom, Absalom!* (1936) provided a map of the fictional Yoknapatawpha County in Mississippi, where several of Faulkner's novels were set; while concurrent to Tolkien's own publications, E. R. Eddison was producing maps for his fantasy works such as the *Zimiamvia* series, Fletcher Pratt included a map in his fantasy novel *The Well of the Unicorn* (1948), Mary Shephard drew a map of Cherry Tree Lane and its surroundings for P. L. Travers' *Mary Poppins in the Park* (1952), and three of C. S. Lewis' Narnia novels, namely *Prince Caspian* (1951), *The Silver Chair* (1953) and *The Horse and His Boy* (1954), depict various areas of Narnia cartographically, as illustrated by Pauline Baynes.

---

[3] The map, complete with compass points and scale, but depicting only a blank space, corresponds to the nonsense lines in the poem: 'He had bought a large map representing the sea, / Without the least vestige of land: / And the crew were much pleased when they found it to be / A map they could all understand' (1982, 683).

This brief overview indicates the rich history of literary cartography that preceded Tolkien, and that provides fertile ground for scholarship on these maps. Surprisingly, however, the critical field is very limited, and largely focuses on fantasy texts.[4] One of the first interventions into this subject was Phillip C. Muehrcke and Juliana O. Muehrcke's 1974 article 'Maps in Literature'. Notable for the breadth rather than depth of their study, Muehrcke and Muehrcke cite a range of fictional maps, as well as references to map-reading, that appear in both fantasy and non-fantasy literature, in order to talk through some of the key characteristics and tensions within cartography, such as the map's pretension to truth versus its invented nature, the map's ability to construct narrative, and the limitations of representation, and to argue that it is these very tensions that make it such a compelling tool for authors. Muehrcke and Muehrcke's study is relatively cursory, yet it is striking for the ways in which it uses literature to comment on cartographic practices and vice versa, treating the maps in the texts as maps rather than illustrative devices.

This approach was largely undeveloped in subsequent studies. Walker's 1981 article 'The Cartography of Fantasy', despite its promising title, dwells largely on the categorization of setting in fantasy literature; his cartographic approach limited to a call for better maps that would 'be a considerable aid in the understanding and enjoyment of a fantasy tale' (1981, 38). Although Walker demonstrates an awareness of and curiosity about the relationship between fictional worlds and their cartographic representations, his argument is largely paratextual: he is interested in the redrawing of maps for fantasy literature in order to ameliorate the reader's experience of the text, rather than in the critical or thematic significances of the map as object. Peter Hunt makes a similar point regarding the map's usefulness for the external reader, arguing that fantasy maps help to structure the narrative and speak to the setting. Hunt does briefly consider the critical value of literary maps: he is interested in 'low fantasy' maps that illustrate the English landscape, arguing that these maps therefore become a tool for engaging with 'landscapes of profound national symbolism' (1987, 13). Nevertheless, just like Walker, Hunt is primarily interested in the map's ability to represent the landscape, rather than in relationships of power between the map, the mapper and the mapped. This former strand is picked up and expanded by Wolf, who examines the map as a world-building tool. As this is the subject of Wolf's study, it is not surprising that he focuses on this particular function

---

[4] Some of these maps have received more scholarly attention, but this is often in the context of their author's broader work rather than of cartography as a literary device.

of the map, yet he limits himself to the map's ability to aid in the author's world-building by representing the world that is described in the text and at times even the world beyond the text, in order to reinforce the reader's belief in this sub-creation. Wolf, however, does not investigate how these maps can contribute to sociocultural and political aspects of world-building, and how the very existence of a cartographic tradition within the sub-creation can add complexity to the narrative.

Deirdre F. Baker does notably engage with the politics of fictional and fantasy maps. Drawing on Jones' satire of the fantasy map, Baker argues that all fantasy maps are effectively reproductions of each other, with each troublingly encoding the same conservative politics within its image. Baker illustrates how these maps, including those in texts by Tolkien, Lewis, Christopher Paolini and Garth Nix, are all structured around socially constructed understandings of cardinal directions: in Tolkien, for example, there is a clear tension between West and East, which Baker argues symbolizes the threat of Nazi Germany and post-First World War I anxieties; while in Lewis, the threat comes from the south of the map from the orientalized land of Calormen, which Baker considers an embodiment of English anxieties surrounding the Muslim world. Although Baker's allegorical readings verge on the reductive, her recognition of the political character of imaginary worlds is important. Unfortunately, Baker does not dwell on how the maps articulate and foster these anxieties, and what this has to say about the practice of cartography, focusing instead on the politics of the text with the map as a mere reflection. Padrón similarly considers the role of the fictional map beyond its engagement with the external reader. In his discussion of the map of More's *Utopia*, he argues '[t]he maps, therefore, are not just maps of an imaginary island, made available so we can see and know it. They are emblems of our desire to know and possess that island ...' (2007, 269–70), also framing the maps in *Gulliver's Travels* as a cartographic expression of the text's satire of travel narratives and their quest for absolute knowledge. Elsewhere, however, Padrón appears to double back on this understanding of the relationship between the text and its map, claiming that '[l]iterature of all kinds has a great deal to tell us about space and place, but the things it has to communicate are not necessarily of the sort that lends itself to cartographic representation. Mapping involves visibility, stasis, hierarchy, and control. Literature often works to subvert these things ...' (2007, 265). Here, Padrón sees the map as a literary device that has limited usefulness, rather than as its own form of expression. Although Padrón's acknowledgement of maps as significant beyond their illustrative value is essential, his brief analysis of each literary text and his lack of engagement with cartographic theory renders this study largely perfunctory.

The most important and rigorous study of fantasy maps is Ekman's *Here Be Dragons: Exploring Fantasy Maps and Settings* (2013). Ekman's approach is what he terms 'topofocal', where 'the setting ... provides a critical way into the work' (2013, 216). Using maps as a touchstone to discuss the significance and complexity of fantasy settings, Ekman acknowledges the political nature of the map, investigating how they are both presented and present within the narrative in order to comment on spatial, temporal and cultural tensions in the texts. Although Ekman is largely interested in contemporary texts from the mid-1970s to the mid-2000s, he draws prolifically upon Tolkien due to, as he argues, his central position within the genre. Beginning with a quantitative analysis of a selection of fantasy texts – which yields conclusions largely concerning the proportion of fantasy literature accompanied by maps and the type of world depicted – Ekman moves onto a more qualitative analysis, which acknowledges both the paratextual and narrative function of the literary map. In particular, Ekman performs a close reading of Tolkien's 'A Part of the Shire' that examines the ways in which the map extends the narrative, by providing information not only about the topography of the Shire but also about how the Shire's inhabitants understand, interpret and represent their home. Ekman touches on certain dynamics of power that the map embodies, arguing that the very act of mapping the Shire, and the level of topographical and toponymical detail it presents, fits into a wider pattern of control over the landscape that is then mirrored within the textual narrative, such as in the agricultural nature of the Shire or the driving back of the Old Forest from the Shire's borders. Ekman then extends this approach to other key issues of setting in fantasy texts, including borders and boundaries, the tension between nature and culture, and political realms and those who rule them. Although these subsequent chapters do not necessarily focus on maps to the same extent as the first chapter, they maintain Ekman's topofocal lens, exploring how various representations of setting can illuminate broader thematic and generic concerns.

Since Ekman's study, there has been little written on literary maps. Maria Sachiko Cecire draws on Foucault's idea of spatial power in order to briefly consider how the map of Narnia contributes to the nationalistic and colonial character of *The Voyage of the Dawn Treader* (1952), arguing that the medieval character of the map 'recalls the political divisions and potential for discovery implied in these earlier images of the world' (2016, 115). A 2018 special issue of *Children's Literature in Education*, meanwhile, examines maps and mapping in children's and young adult literature, engaging with the broad spectrum of paratextual, political and technological possibilities of literary cartography. Indeed, the most important recent intervention in the field is Ekman himself: drawing on cartographic theory by Wood, Fels and Matthew H. Edney, which argues that every map has an author, a subject and a theme, Ekman reads the map of Brandon Sanderson's *The Rithmatist* (2013) in order to discover the map's potential for illustrating the fantasy world's conflicts, and in particular its colonial tensions. Recent scholarship, scant as it is, is therefore turning towards the fantasy maps' political and narrative possibilities and away from the map as mere paratext or world-building device. Yet this approach still remains limited: there is yet to be an extensive study that engages, as Cooper argued, with critical cartography that centres not only on a map but a corpus of maps at the centre of the discourse in order to demonstrate the pervasively political nature of cartography and its intersections with the politics of the text.

Structure and Overview

Using the critical cartographic framework outlined above, this book will examine how cartography in Middle-earth intersects with, expresses and facilitates exercises of power, understood through a Foucauldian lens. The first chapter examines the ways in which real-world cartography has historically been bound up in facilitating power, focusing on both medieval and modern traditions to demonstrate the map's ability to encode power, politics and ideology, that is to say, systems of thought informed and constructed by particular political persuasions, which are communicated and crucially promoted through various social and cultural apparatuses – including, in this case, cartography. By framing cartography as historically inextricable from power relations, it becomes striking how Tolkien draws on particular stylistic and conceptual cartographic traditions from these two periods that enable such enmeshments of power, thereby creating a fictional cartographic tradition that is aesthetically and structurally informed by cartography's historic relationship with power.

The second chapter turns to a diegetic literary analysis that continues throughout the rest of the book. This chapter examines the relationship between humans and the environment, positioning cartography as a tool of the human/nature binary that represents the control that humans attempt to enact over the natural world. Drawing on the work of ecocritics and posthumanists such as Val Plumwood and Donna Haraway, this chapter positions Tolkien's writing as a form of environmental protest that mirrors some of the key positions of modern ecological thought, by seeking to break the hierarchy between human and nature that the binary creates and framing the non-human as an actant with its own complexity and agency. In doing so, I want to place Tolkien's engagement with environmental concerns within a contemporary context of the emerging ecological crisis and as an author writing into the beginnings of the Anthropocene.

The third chapter continues this emphasis on situating Tolkien within his own period, and specifically within its attitude to time. The chapter examines the ways in which Tolkien's cartography, and cartography more broadly, is capable of mapping time and temporalities, and in particular the changing social and political conceptualizations of time that were taking place. By outlining the shift in understandings of temporality due to eighteenth- and nineteenth-century scientific discoveries of deep time, uniformitarianism and evolution through the writings of Charles Lyell, James Hutton and ecocritics such as David Farrier and Jeffrey Jerome Cohen, I establish Middle-earth's cartography as articulating the tension between human and non-human timescales, as well as anxieties around these timescales and the dislocation of the human within the passage of time.

The fourth and final chapter examines the ways in which cartography is part of broader structures of power surrounding politics and conflict over land, from smaller-scale, individualized conflicts to aggressions formed by broader imperialist movements. Drawing on postcolonial theory by critics such as Frantz Fanon and Edward Said, as well as postcolonial ecocriticism by critics such as Elizabeth M. DeLoughrey, George B. Handley, Graham Huggan and Helen Tiffin, this chapter unpacks how conflict over land is used to perpetuate hierarchies of power, the aligning of the non-human with the considered-inhuman that lies at the heart of environmental racism and the racialization tactics of imperialism, and the ways in which cartography reflects and participates in this simultaneous exploitation of the land and its people.

Through these investigations, this book demonstrates both the methodological need to study literary cartography within a critical cartographic framework – that takes into account the ways in which maps work in a post-representational sense, and that fully embraces the complexities, nuances and subtexts of the cartographic project – and the literary imperative to place Tolkien's works within these theoretical contexts, thereby opening his writing up to critical evaluations that have largely been neglected. There exists an overwhelming anxiety in the field of scholarship that Tolkien is not taken seriously as a subject of study; that he is not considered within the same calibre as other writers of his period.

The solution to this, I argue, can only be active participation in contemporary literary discourse that demonstrates and validates Tolkien's applicability across theoretical fields, and positions his writing as both political and politicized – a text that sought to reflect the conditions of his own world as well as to create new ones.

1

Political Map-making

In *Sylvie and Bruno Concluded* (1893), the second volume of his unsuccessful follow-up to *Alice's Adventures in Wonderland* (1865), Lewis Carroll briefly touches on the subject of cartographic representation through Mein Herr, a traveller from an unspecified land who regales the eponymous fairy duo with tales of everyday life from his home country. Discovering a pocket map among his belongings, Mein Herr asks Sylvie and Bruno about the largest map that they would consider useful, prompting this exchange:

> 'About six inches to the mile.'
>
> 'Only six inches!' exclaimed Mein Herr. 'We very soon got to six yards to the mile. Then we tried a hundred yards to the mile. And then came the grandest idea of all! We actually made a map of the country, on the scale of a mile to the mile!'
>
> 'Have you used it much?' I enquired.
>
> 'It has never been spread out, yet,' said Mein Herr: 'the farmers objected: they said it would cover the whole country, and shut out the sunlight! So we now use the country itself, as its own map, and I assure you it does nearly as well ...'
>
> (Carroll 1982, 556–57)

Mein Herr's nonsensical conceptualization of cartographic accuracy prefigures the 1946 short story by Jorge Luis Borges entitled 'On Exactitude in Science', a tale which similarly envisages maps at such a large scale that 'a single Province occupied the entirety of a City, and the map of the Empire, the entirety of a Province ...' until, tiring of this supposed inaccuracy, the Cartographers Guild creates 'a Map of the Empire whose size was that of the Empire, and which coincided point for point with it' (2018, 35). Future generations eventually deem the map useless, so that it disintegrates into tatters that are strewn throughout the empire.

There is on the surface very little that unites Carroll's Victorian nonsense tradition with Borges' Argentinian avant-garde, but both of their absurdist takes on cartography are striking in the broader questions they open up about a map's representational capacities. For both Mein Herr's compatriots and the members of Borges' Cartographers Guild, the only way to ensure perfect cartographic representation is to create a one-to-one scale map of the land; a project that ultimately fails in desirability and functionality. The stories push the idea of a representational map – and perhaps, more broadly, the very idea of representation – to a farcical extreme in order to lay bare its absurd propositions: that any human-led curation of the world might contain all of its complexities and accuracies.

The encoding of social and political subjectivities in cartography is not necessarily synonymous with a technically inaccurate map; however, it similarly demonstrates the innate tension in cartography as both factual representation and artificially constructed object. Denis Wood comments on the mythic concept of the entirely objective map, arguing that the vocabulary used to describe maps – 'mirrors', 'accurate', 'neutral' – works to disguise the map as a socially constructed reproduction rather than representation of the world. '[I]s any myth among cartographers more cherished,' Wood asks, 'than that of this map's dispassionate neutrality?' (1992, 22). This dispassionate neutrality is also contested by J. B. Harley, whose work – as discussed in the Introduction – positions maps as far removed from Carroll's and Borges' fictional, purely representational documents, and instead as explicitly political and politicized objects. Taking an iconological approach, Harley argues that maps are value-laden images; both in terms of what they choose to omit and represent and the systems of signs and styles that they employ, maps are a way of 'conceiving, articulating, and structuring the human world which is biased towards, promoted by, and exerts influence upon particular sets of social relations' (1988, 278). Harley reconceptualizes maps as 'literary' texts in order to draw attention to issues of sociopolitical context and the ways in which maps reciprocate these world views through their iconography, as well as factors of readership, authorship and carto-literacy. Ultimately, Harley argues, maps are a vehicle for knowledge; not the objective factuality that they often purport to express, but for systems of knowledge that are used to maintain particular power relations, which are reinforced through cartographic symbolism.

This chapter assesses Tolkien's cartography from this perspective, that is to say, not as an accurate representation of his fictional landscapes, but as a socially and politically constructed text that imbibes and reflects the sociopolitical conditions of its production in both the primary and secondary world. While the following three chapters examine this idea from particular theoretical angles in order to consider how maps, map-making and map-reading in Middle-earth are profoundly politicized concepts, this chapter will take a more extradiegetic approach, highlighting cartographic representational methodologies that Tolkien drew on from the primary world that are themselves explicitly ideological. This chapter is therefore less interested in the fictional sociopolitical conditions that can be read through the maps and that are discussed at length in the following three chapters, but rather in the ways in which Tolkien uses politically embedded methods from historic practices in order to situate his cartography within a tradition of subjective map-making and value-laden image-making. This deliberate positioning then enables the further enmeshments with political and ideological narratives that are explored in the following chapters.

As comparing Tolkien's maps to the breadth of cartographic history would be unfeasible within the limits of this book, I will focus on two distinct periods: the medieval and the modern, the latter defined as post-Enlightenment through to Tolkien's own time. This is not to argue that other genres and periods of map-making would not have shaped Tolkien's work, but as a limited exemplar of historic influence, the period that he was directly drawing from – the medieval – and the period he was writing within – the modern – are arguably the most relevant. The medieval influence on his cartography is indeed already critically established; much as Tolkien's texts have typically been read through a medievalist lens, Tolkien's maps are also widely considered medievalist artefacts. Ricardo Padrón argues that although Tolkien's maps gesture towards contemporary techniques through the use of scales and a compass, 'on the whole they resist the abstraction of modern cartography, preferring a deliberate, stylized archaism that echoes Tolkien's writing ...', embodied through the iconographic depiction of natural features, rather than the use of abstract symbols, and the 'vaguely old-fashioned' typography of the map (2007, 273). Padrón's argument is typical of the admittedly limited source studies that have been undertaken into Tolkien's cartography, in that he references the 'archaic' nature of the maps without situating them within a specific cartographic practice. Similarly, Dimitra Fimi discusses the 'typically anachronistic, medieval way' the I Vene Kemen map in *The Book of Lost Tales* depicts both the Two Trees and the Sun and the Moon (2009, 124); Karen Wynn Fonstad explains that 'Tolkien was envisioning a world much as our medieval cartographers viewed our own …' by portraying the world as a disk (1994, ix); and Christina Scull and Wayne G. Hammond argue that the appearance of the Lonely Mountain in Thror's Map is very like the look of mountains on certain medieval woodcut maps (1995, 94).

Although each of these studies supports the characterization of Tolkien's cartography as 'archaic', with most indicating a specifically medieval influence, the comparison between the source material and Tolkien's own attempts remains somewhat cursory. Medieval understandings of geography, place and politics, the ways in which these were expressed through maps and the development of typical cartographic practices throughout the medieval period are not commented on, so that we are left with a vague but generally accepted categorization of Tolkien's maps as 'medieval'. The exception to this is Jason Fisher, who sources Tolkien's 'Circles of the World' within a medieval tradition of depicting the world as a flat circle, as seen in the *Heimskringla*, the Latin Vulgate Bible, and in particular in the Hereford Mappa Mundi. Fisher explores both stylistic similarities between Tolkien's maps and the Hereford Mappa Mundi – such as the orientation of the world to the east and the encircling sea found in both Tolkien's 'Ambarkanta' maps and the Hereford Mappa Mundi – as well as conceptual similarities – in particular, the use of the limited, bound circle to reinforce the connection between mortality and the physical world (2010, 11).

Although Fisher's argument is persuasive, the broader field that continues to categorize Tolkien's cartography as simply medieval neglects the elements of his cartographic corpus that complicate this characterization, and that are influenced by more contemporary practices. Tolkien was certainly familiar with modern maps, arguably more so than their medieval counterparts: in his essay 'A Secret Vice', Tolkien describes a moment during training in the First World War, when he was sitting in a tent 'listening to somebody lecturing on map-reading …' (1997a, 199). Map-reading was a key part of Tolkien's war experience: his early training in the Officers' Training Corps at Oxford involved one lecture a week and classes in signalling and map-reading on free afternoons (Hammond and Scull 2006, 55); he studied from the book *Signalling: Morse, Semaphore, Station Work, Despatch Riding, Telephone Cables, Map Reading*, edited by E. J. Solano (1914) (ibid., 72); and he eventually chose to specialize in these skills, obtaining his Provisional Instructor's Certificate of Signalling (For Officers) on 13 May 1916, with a 95% accuracy on Written Examination, Examination of Telephony, and Knowledge of Map Reading (ibid., 80).

Aside from his expertise in military map-reading, Tolkien almost certainly also used maps in his everyday life. In particular, his fondness for walking in the countryside[1] probably led to a familiarity with large-scale topographical maps such as the Ordnance Survey. Although Tolkien never references using such maps as the Ordnance Survey on his walking trips, it is probable that in lengthy trips in unfamiliar places, such as his journey through the Swiss Alps where his company purposefully avoided main roads, or his family holidays to Cornwall, he would have used such detailed map for orientation. Moreover, many of Tolkien's walking trips took place with his friend C. S. Lewis, who was famously passionate about walking, taking an annual walking tour with friends which would last several days. Although there are no records of Lewis using maps on these tours either, a letter Lewis wrote to his illustrator Pauline Baynes regarding the maps for his Narnia books details that '[m]y idea was that the map should be more like a medieval map than an Ordnance Survey …' (Cecire 2016, 115).

The combination of medieval and modern influences in Tolkien's cartography is uniquely remarked upon by Stefan Ekman, who reads it through Umberto Eco's theory of the pseudomedieval, as discussed in *Travels in Hyperreality* (1986). Eco examines the way in which contemporary culture is infused with attempts to replicate and simulate reality, arguing that industries and technologies as diverse as Disney, holograms and Superman work to create what Eco terms hyperreality, an artificial reality which is more detailed and tangible than actuality, so that the simulacrum comes to replace the original. Eco applies this concept to contemporary culture's relationship with the medieval, arguing that while there has been a 'return to the Middle Ages' in the modern period (1987, 65), this interest oscillates between 'fantastic neomedievalism and responsible philological examination' (1987, 63). Eco claims that most literary and artistic products fall into the former category, reconstructing a version of the Middle Ages that is fictional, yet which is often read as authentic. Eco stresses that these cultural products adapt the Middle Ages in order to 'meet the vital requirements of different periods', using them as a 'mythological stage' on which to project contemporary ideas (1987, 68). Ekman posits that although fantasy maps as a genre nod to medieval practices through certain stylistic

---

[1] In the summer of 1911, Tolkien, along with his brother Hilary and his aunt Jane Neave, joined a walking tour in the Swiss Alps organized by some family friends, where they hiked along mountain paths, avoiding the main roads (Hammond and Scull 2006, 27). The next year, during the summer vacation of 1912, Tolkien went walking in Berkshire, recording the scenery in his sketchbook (2006, 34). In August 1914, Tolkien 'explores the Lizard Peninsula in Cornwall on foot with Father Vincent Reade' (2006, 53). In the summer of 1932, Tolkien and his family go on holiday to Cornwall, taking long walks to Land's End (2006, 164). Tolkien also went walking with Lewis, on one occasion accompanying Lewis and Owen Barfield on a walking holiday in the Quantock Hills in Somerset in April 1937 (2006, 194).

features, actual medieval techniques are largely simplified and combined with modern techniques in order to create a medieval impression while still being recognizably 'map-like'.

This chapter intends to nuance the reading of Tolkien's maps as vaguely medievalist by examining how his cartographic practice also draws from modern methodologies. Anticipating the following chapters, which consider how Tolkien's maps and narratives respond to contemporary sociopolitical issues, this integration of the modern with the medieval will realize Eco's argument that the medieval is employed as a stage upon which contemporary politics and anxieties can be projected, an extension of Tolkien's broader textual strategy of using medievalist structures and imagery to address distinctly modern concerns. In doing so, this chapter will demonstrate how Tolkien drew upon particular political and ideological techniques, both medieval and modern, in order to create a tradition of map-making that is not merely representational or reproductive, but rather constructive: one that denies the neutrality of maps, and opens itself up to their sociopolitical conditions. Beginning with a historical overview of medieval and modern cartography, this chapter will tease out the influences in Tolkien's cartographies through a close reading of a limited selection of Tolkien's cartographic corpus in order to illustrate how these mimic the primary world's political and ideological encodings. It is through tracing the maps' textual and extradiegetic production histories that we can come to understand how Tolkien shaped each map not as a one-to-one representation of his world-building, but rather as a vehicle for sociopolitical and ideological expression.

Medieval Cartographic Practices

Map-making did not originate in the medieval period. The desire to understand and interpret one's surroundings, and thereby position oneself in the world, has existed since the prehistoric period,[2] and continued to develop throughout the ancient world and classical antiquity. Although, as P. D. A. Harvey argues,

---

[2] Catherine Delano Smith unequivocally states:
> There is no doubt that by the beginning of the Upper Paleolithic [c. 40,000 BC] man possessed both the cognitive capacity and the manipulative skills to translate mental spatial images into permanently visible images. It is possible to identify alternative modes of cartographic expression in the rock art record, ranging from the supermundane to the real world, for instance, and including perceptions of landscape from sometimes a low, sometimes a high, and occasionally, a vertical angle. (Delano Smith 1987, 62)

For further discussion of prehistoric cartography, see Delano Smith's full chapter.

the relationship between medieval cartography and its preceding model from antiquity is at times difficult to establish[3] (Harvey, 1987b, 283), medieval cartography was nevertheless emphatically the continuation of a longstanding tradition. European medieval cartography itself is defined as those maps produced from the post-Roman fifth century through to the pre-Renaissance fifteenth century. It did not, however, remain static throughout this millennium, but rather developed continuously, engendering numerous distinct cartographic traditions and practices according to specific periods and places. As this overview of medieval cartography will act as a contextualization of Tolkien's maps, it will focus on cartographic production in medieval England, with reference to wider European trends when relevant.

Broadly, English medieval cartography was informed by popular European conceptualizations of the world and spatial representation. David Woodward separates medieval European cartography into three key periods, each roughly aligning with the start of a new medieval renaissance (1987, 299). The first period begins in the fifth century and lasts until the end of the seventh century, and is characterized by three key cartographic traditions, named after the authors who popularized them: Macrobius (c. AD 395–436), Orosius (c. AD 383–post 417) and Isidore (c. AD 560–636). The maps produced in these traditions were mappae mundi, or world maps. The Macrobius map depicted the Earth split into five climatic zones: a polar zone located in the north and in the south, a central equatorial zone, and two temperate zones, each sandwiched between the polar and equatorial zones. The temperate zones are considered the only habitable zones, and it is mainly here that the known continents of Europe, Asia, Africa and the Antipodes were located (1987, 300). The Orosian and Isidorian models took a less zonal and more geographic approach: known as tripartite, or T-O maps, they separated the world into three continents, with Asia located in the east (at the top of the map), bordered by both Europe and Africa and encircled by an ocean (Williams 1997, 17). All three of these map forms had a profound impact on medieval cartography beyond their inception in this first period, with examples of their influence found in maps dating through to the Renaissance (Woodward 1987, 299).

The second period stretches from the beginning of the eighth to the end of the eleventh century. Thanks to the marked increase in production of manuscripts

---

[3] For further discussion of the link between ancient and medieval cartography, see Richard J. A. Talbert and Richard W. Unger's edited collection *Cartography in Antiquity and the Middle Ages: Fresh Perspectives, New Methods* (Brill, 2008).

and texts for cathedrals and monastery schools during the Carolingian Renaissance, the maps of this period are characterized by a pronounced religious influence, leading to the period being termed 'the golden age of Church cartography' (Woodward 1987, 299). Among the key maps produced in this period are the Beatus maps. These maps accompanied the text of Beatus of Liébana's *Commentary on the Apocalypse* (eighth century AD). The extant maps are closely based on Beatus' original manuscript map, which has now been lost (Woodward 1987, 303). These maps are very like the tripartite model popularized by Orosius and Isidore, with one notable exception: there is the addition of a fourth continent, called the Antipodes. The purpose of these maps was to illustrate the journeys of the apostles as they went forth into the world to evangelize the word of Christ (Riffenburgh 2014, 24).

The third period is relatively short, beginning in the twelfth century and ending in the final years of the thirteenth. This period was marked by a renewed influx of geographical and cosmological knowledge, as several previously untranslated Arabic and Greek texts – including Ptolemy's *Algamest* – were made available to readers of Latin. It was also at this time that Europe saw the emergence of its first universities: Bologna, Oxford and Paris were all founded roughly in the twelfth century, and the subjects which they taught – mathematics, astronomy and geometry – were all directly linked to cartography and to a wider interest in the physical world (Woodward 1987, 306). This period saw the creation of many of the most famous mappae mundi, including the Ebstorf map (*c.* 1235), the Psalter map (thirteenth century; British Library Add. 28681), the Hereford map (*c.* 1300) and the Sawley map (*c.* 1190–1210; Corpus Christi College, Cambridge MS 66), which in form were based on the previous Orosian and Isidorian models. The maps in this group show a shift away from maps as manuscript illustrations and instead become independent documents; both the Ebstorf map and the Hereford map are sizeable artefacts that would have been hung up and displayed.

Although the zonal model of Macrobius and the tripartite models of Orosius and Isidore were based on their respective texts' assertions about the structure of the physical world, mappae mundi were nevertheless not accurate diagrams of the Earth or its landscape, and, moreover, were never intended to be such. George H. T. Kimble discusses the false perception of medieval maps as vehicles for geographical accuracy, explaining that the contemporary purpose of the maps was as works of artistic rather than informative presentation: their authors 'would have branded any man a fool who might have supposed that he could determine the distance from London to Jerusalem by putting a ruler across a map ...' (qtd in Turchi 2004, 34–35). As Harvey argues, it was not only limitations of technique that resulted in this lack of accuracy (although these certainly existed) but also definitions of concept (1991, 7). As indicated by the Golden Age of church cartography and the presence of maps such as the Psalter map in religious volumes, the purpose of mappae mundi was underpinned by contemporary theological and sociopolitical concerns: mappae mundi were intended to convey not geographical information, but rather religious narrative. Together, the symbols and structures on these maps worked to convey an ideological interpretation of the world, one rooted in the key ideology of the European Middle Ages: Christianity. An approximate depiction of the world is provided in order to comment on and teach the narrative of the Christian experience within the whole of God's creation, and to embody 'the Christian world view of a divine order' (Riffenburgh 2014, 22).

Mappae mundi encoded these narratives in multiple ways. Woodward argues that the maps act as a compendium of the three main stages of the Christian narrative: Creation, the Passion of Christ and the Last Judgement. The Creation is symbolized through the three continents in the tripartite model: according to the Bible, each continent was peopled by one of Noah's three sons, making the tripartite model a reflection of the historic beginnings of humankind (Woodward 1987, 334). The Creation story can also be seen in the representation of the Garden of Eden in many of the mappae mundi, including the Hereford map, the Ebstorf map and the Psalter map. The Passion of the Christ is represented in the T-O layout of the tripartite model, with the T representing the cross Jesus died on (1987, 334). Finally, the Last Judgement is seen in the presence of 'Christ in Glory' featured at the head of or encircling the world, indicating His ultimate jurisdiction over the physical earth (1987, 335). This is seen in the Psalter map, where Jesus is depicted at the top of the world, flanked by two angels, and with his hands held out in a blessing; the Ebstorf map depicts Jesus' head at the top of the world, his hands either side and his feet at the bottom, thereby depicting him as both encompassing and integrated within the world.

Mappae mundi also embedded Christian ideology in their very structure. Many medieval maps placed East at the top rather than North, as these maps oriented themselves towards the Middle East and the presumed location of the Garden of Eden. In the Hereford, Psalter and Ebstorf maps, the Garden of Eden is represented at the very top of the map through the iconography of Adam and Eve and the forbidden tree. The story of creation, and more broadly God's presence in the narrative of the world, is immediately prioritized through the map's orientation. Other key biblical locations also affected the structuring of the mappae mundi: Denis Cosgrove describes medieval world maps as 'center-enhancing' (2007, 79), in that they placed what was considered significant and valuable in the middle of the map; frequently, particularly in the later Middle Ages, this meant centring the map on Jerusalem – the source of Christianity.[4] This practice is seen in the Hereford map, the Ebstorf map and the Psalter map, which all position Jerusalem as the centre of the world, both physically and metaphorically. Moreover, in the Ebstorf map, the city of Jerusalem takes up as much surface area as the entirety of Britain, reinforcing the map's emphasis on theological and sociopolitical representation rather than geographical accuracy.

Naturally, when the centre of the map is encoded with meaning, so too are the edges. In mappae mundi, the further from the centre and towards the edge the reader travels, the more dangerous and frightening the territory becomes. The safety of the known world is juxtaposed with edges populated by cannibalistic tribes, monsters and natural disasters. In the Ebstorf map, animals become more fantastic towards the edges, with griffins and demon-like creatures portrayed. There are also depictions of violent murders and humans eating human body parts, which imbue the edges of the map with taboo and danger. In the Hereford and Psalter maps, depictions of monstrosity are similarly rife, with mythical races such as sciopods (one-legged men) and blemmyes (headless men) populating the edges. Medieval cartographers responded to the unknown, blank spaces and far away edges of the world with uncertainty and fear, using their maps as a form of commentary on what was considered knowable, and therefore safe. This cartographic technique speaks to broader medieval codifications of space. Examining how manuscript pages similarly fill their margins with strange creatures, Michael Camille emphasizes that politics of space and representation were engrained within medieval art and iconography, arguing that medieval people were 'highly sensitive to disorder and displacement precisely because they were so concerned with the hierarchy that defined their position in the universe' (1992, 16), channelling this concern with disorder into visual artefacts

---

[4] Not all maps produced throughout the Middle Ages were centred on Jerusalem; Woodward points out that in the early Middle Ages, very few were. However, as Woodward notes: 'The strengthening of the idea of Jerusalem as the spiritual center, a natural outcome of the Crusades, may have been responsible for a noticeable shift in the structure of mappae mundi from 1100 to 1300 ... [so that] the practice of placing Jerusalem at the center became common' (Woodward 1987, 342).

such as illuminated manuscripts and mappae mundi. These artefacts subverted social order in their margins, using the limits of the page to signify the limits of representation; in doing so, the edges worked to highlight the symbolism of the normative by acting as its foil, reinforcing its central position both on the page and in society.

Woodward's analysis of the trajectory of medieval cartography largely centres on world maps; however, it is important to note that other forms of cartography were present – if not as prolific – in the Middle Ages. Portolan charts began to emerge in the thirteenth century and were a result of increased marine exploration, thus concentrating on coastal outlines rather than portraying any inland features, unlike the mappae mundi, which mainly focused on terrestrial features. The charts are characterized by intricate depictions of coastlines and criss-crossed rhumb lines for navigational purposes, created by marking sixteen equidistant points (among them the main cardinal points) along a circle that extended over the map; these points would then be joined up with intersecting lines (Harvey 1991, 43). Given the maritime nature of portolan charts, their production was largely concentrated in Mediterranean port towns. Although it is difficult to trace the origins of the very first portolan charts, the earlier examples from the first half of the fourteenth century were produced in the cities of Palma, Genoa and Venice. By the second half of the century, the charts can, broadly speaking, be divided into two main groups: the Italian charts and the Spanish charts, with each tradition exhibiting its own uniquely identifiable characteristics and features.

What sets portolan charts apart from other medieval cartography is their representation of and adherence to scale and direction, concepts which, as demonstrated above, were largely irrelevant to mappae mundi. Tony Campbell states that portolan charts 'were the most geographically realistic maps of their time' (1987, 445), making them the closest relative to modern topographical maps, which strive for spatial accuracy rather than the topological symbolism of mappae mundi. The grounds for this accuracy are rooted in the function of portolan charts; these are maps used primarily for navigational purposes, making precise, to-scale representations of the physical landscape of paramount importance. The style of portolan charts is largely defined by this function. As the charts were used primarily for seafaring purposes, it is the coast that is depicted in the highest detail, with little to no terrestrial detail featured on the majority of charts. The charts' rhumb lines are another characteristic feature derived from their navigational function. Although rhumb lines acted as a compass, from relatively early on in their history, portolan charts featured an additional compass rose alongside the lines,[5] as well as a scale bar; cartographic features that would become a fundamental part of future topographic and navigational maps (Harvey 1991, 48).

Overall, then, portolan charts contain considerably less subjectivity and symbolism than the mappae mundi. This is not to say, however, that portolan charts were free of sociopolitical and cultural markers – no map, as Carroll and Borges told us, are ever free from the politics of representation. The Spanish Dalorto chart of 1325 (Archivio del Palazzo Corsini, YYef 2014-561) and the Dulcert chart of 1339, both drawn by the same cartographer, Angelino Dulcert, feature a relatively large amount of inland detail compared to Italian charts of the same period; mountain ranges are picked out in striking green and blue, zigzagging rivers cross lakes and towns, and non-coastal towns are marked and named, often with a flag or emblem representing the kingdom or settlement. Campbell argues that these maps are more than ornamental, they symbolically signify the ruling dynasty in the area (1987, 393), acting as – to use Harley's term – value-laden images, conveying information that is superfluous to their navigational function.

The proliferation of portolan charts coincided with the emergence of regional and local maps. This genre of map was relatively rare in the Middle Ages; the first regional maps did not appear until the mid-twelfth century, and even then, their production was limited. These rare maps can be divided into two categories: local maps of very small areas, and regional maps that cover entire countries. Of the former, very few survive from medieval England: from the mid-twelfth to the mid-fourteenth century there are only three extant examples, from the mid-fourteenth century to the beginning of the sixteenth century approximately thirty, and from the first half of the sixteenth century approximately two hundred (Harvey 1987a, 464). These maps vary extensively in what they portray; they include an 1150 plan of Canterbury Cathedral (Trinity College, Cambridge MS R.17.1) – the earliest large-scale English map still existing – 'strips in arable fields, house plots in towns, rivers with mills and fisheries … [and] whole tracts of countryside including towns and villages' (1987a, 484).

Concerning broader regional maps, very few that both depict and were produced in England remain. The most well-known examples are the four

---

[5] The Catalan atlas of 1375 (Bibliothèque nationale MSS. Esp. 30) has a miniature compass rose on its left-hand edge; this stylistic feature became normal in subsequent portolan charts of the fifteenth century.

extant maps of Britain drawn by Matthew Paris and the Gough Map of Britain. Matthew Paris' maps are found in various St Albans manuscripts and were drawn in the mid-thirteenth century,[6] and have been categorized as the 'first truly regional maps and the first detailed maps of Britain ...' (Connolly 2009, 186). Each map shows an outline of Britain with a network of cities and rivers spanning across it, constructed around a vertical axis which runs from the northern town of Newcastle to the port town of Dover in the South of England. Suzanne Lewis has described Matthew Paris' Britain maps as 'a genuine attempt at making a map in the modern sense of the word ...' (1987, 365) in that they prioritize geographical representation above theological symbolism, being oriented towards the north rather than the traditional east,[7] and focusing on physical features such as cities and rivers. Stylistically, the maps present the same pictorial quality as the smaller, local maps of the period: Hadrian's wall is depicted as an actual, turreted wall, while cities and administrative centres such as London are portrayed as small castles. Produced roughly a century later in the mid-fourteenth century, the Gough Map (Bodleian Library MS. Gough Gen. Top. 16) is similar to Matthew Paris' maps, in that it shows an outline of Britain with a network of cities, rivers and roads. Unlike Matthew Paris' maps, however, which are effectively an itinerary from Newcastle to Dover, with other cities and landmarks built around this predominant axis, the Gough Map does not focus on just one route, but rather shows an entire network of roads around Britain, situated correctly between an elaborately drawn river system, with a small figure by each section of road giving its length in local miles (Riffenburgh 2014, 73).

Local and regional cartography was thus a broad and largely undefined genre in the Middle Ages, encompassing numerous geographical modes of representation. It is the regional maps, however, that come closest to a modern navigational function. Both Matthew Paris' maps and the Gough Map are based on itineraries and portray ways of getting from point A to point B. The former are based around a single itinerary, while the Gough Map takes a more complex approach of combining numerous itineraries so that the reader

---

[6] The first map (British Library Royal MS 14 C VII), an incomplete copy, is found in a copy of Matthew Paris' *Historia Anglorum*, a history of the English. The second map (British Library Cotton MS Claudius D VI) is the most elaborate and prefaces Matthew Paris' *Abbreviato chronicorum Angliae*, a short chronicle of English history; a third (British Library Corpus Christi MS 16) accompanies a copy of his masterwork the *Chronica Majora*; a fourth, rougher version (British Library Cotton MS Julius VII) exists in a manuscript of various St Albans productions (Mitchell 1933, 27–28).

[7] Several critics, among them Katharine Breen, have pointed out that this orientation may not have been a choice based on scientific accuracy, but rather based on the restrictions of the physical page of his codex: 'Britain, being taller than it is wide ... fit[s] within the traditional rectangular format of his manuscript codex ... ' (2005, 59–60). Although this may well be the case, this still demonstrates an unconcern for the theological implications of the traditional eastern orientation of medieval maps.

can trace their own route through the country, yet they largely fulfil the same function. These maps are emblematic of a pivotal moment in the history of cartography; throughout most of the Middle Ages, this sort of information would be set out as a written list of instructions or descriptions (Harvey 1987a, 464). In the mid-medieval period, these itineraries began to be visualized; the most famous of these visualizations were drawn by Matthew Paris, and feature in the *Chronica Majora* (British Library Royal MS 14 C VII), the same manuscript which contains one of his Britain maps. These itinerary maps are a series of strips spread over seven pages outlining the road between London and the holy pilgrimage sites of Rome and Jerusalem through step-by-step illustrated instructions from one city to the next. The map asks the reader to follow the route up and down the page, similar to a vertical comic strip, thereby allowing them to 'handle, manipulate, and trace their motions across the surface of the world …' (Connolly 2009, 6). What Matthew Paris' Britain maps and the Gough Map achieve, then, is a reconceptualization of this itinerary in the form of the British Isles. It reveals a shift in attitudes towards navigation as a function of cartography: demonstrating how these routes fit into the physical space of the landscape becomes an integral practice, rather than viewing the routes and the space they cover as two unrelated concepts. Although these maps almost certainly were not used for navigation in the way we would use a modern map, in that they were not brought along on travels, they nevertheless are amongst the first to both conceive of Britain as a navigable entity, and maps as capable of conveying such information.

Moreover, due to their itinerary function, both Matthew Paris' maps and the Gough Map introduce the concept of scale, while never quite perfecting it. The first of Matthew Paris' maps includes a note that explains that the island should have been drawn longer, if the page would have allowed it (Lewis 1987, 365). The Gough Map, meanwhile, notes distances between settlements in local (and therefore unstandardized) miles. Although the depiction of the network is therefore not to scale, the map attempts to engage with both a spatial and temporal representation. As Harvey argues, from this it is conceptually 'a very small step to set out the itinerary with its distances all in due proportion' (Harvey 1987a, 496). This new interest in geographical representation lacks the symbolic potency of the theological mappae mundi that made them such distinctly value-laden images, yet it nevertheless embeds a particular sociopolitical context, speaking to a gradually reconfiguring relationship between people and their environment and the increasing human impulse to render the world both navigable and controllable.

Modern Cartographic Practices

This objective of bringing the external world under its totalizing control has been fully realized in the emergence of modern cartography. Although in terms of iconography, cartography underwent significant changes from the medieval to the modern period – a cursory comparison of medieval and post-Enlightenment maps reveals a shift from illustrative and figurative depictions of the world to a more abstract system of signage – modern maps nevertheless remain value-laden images, visually encoding and enabling particular ideologies through these signs. This shift to the abstract, and the use of simplified and regulated symbols rather than individualized pictures to depict the mapped subject, also represents the key change in cartography's purpose from the medieval to the modern: no longer concerned with conceptual narratives or theological contemplation, maps instead purportedly aim to accurately represent and reproduce the world, collecting, organizing and disseminating qualitative and quantitative information about topography and natural and urban environments. This attempt at a comprehensive reproduction of the landscape, pointedly ridiculed by Carroll and Borges, indicates a new value system based on totalizing control through depiction rather than Christian ideals.

This emphasis on representational accuracy did not first emerge in the modern period – its roots can be traced back to the portolan charts and regional maps of the Middle Ages – yet it became crystallized as the primary objective of map-making during the Enlightenment. This was bound up in mutually informing practical and ideological factors. On the practical side, the technological developments of the period enabled cartography's aspiration towards total objectivity. Before the eighteenth century, map-making was a vague art; map-makers tended to create new maps based on amalgamations of previous maps, rather than conducting new surveys of the land, which resulted in new maps replicating previous errors (Hewitt 2010, xxii). Moreover, even when map-makers did conduct their own surveys, their instruments were not yet advanced enough to produce the kind of accuracy seen in modern topographical maps. The telescope was not invented until the beginning of the seventeenth century, was not used in surveying until 1670, and from then until the mid-eighteenth century, problems with the lenses would produce inconsistent results. Other surveying instruments were also prone to inaccuracies, often expanding or shrinking due to temperature differences; chronometers and clocks would speed up or slow down; and the measuring scales that produced these instruments were not technologically advanced enough to calculate minute differences in weight, meaning the same instrument would often produce different results (Hewitt 2010, xxii). By the eighteenth century, however, dramatic improvements in technology meant that map-makers could produce maps with a hitherto unachievable level of accuracy, so that by the second half of the century, 'Britain was home to some of the most precise map making and astronomical instruments in the world and the most diligent, rational surveyors ...' (Hewitt 2010, 4).

Rachel Hewitt's description of 'precise' map-making and 'rational' surveyors points to the Enlightenment ideologies that underpin this new cartographic objective. Developing symbiotically alongside the scientific and technological discoveries of the period, the philosophies of the Enlightenment became integral to the fundamentals of modern cartography and continue to inform the perception of map-making and usage to this day. Harley argues that

> [a]lthough cartographers write about the art as well as the science of mapmaking, science has overshadowed the competition between the two. The corollary is that when historians assess maps their interpretation is moulded by this idea of what maps are supposed to be. In our own Western culture, at least since the Enlightenment, cartography has been defined as a factual science. The premise is that a map should offer a transparent window on the world.
>
> (1990, 3–4)

Wood and Fels explain that Enlightenment cartographers believed that 'a mirror of nature can be projected through geometry and measurement ...' (2008, 6), while Jeremy Black frames the emphasis on accuracy as an 'ideology' all of its own, arguing that cartographic accuracy is largely seen as a by-product of objectivity, a confirmation of the cartography as a science: 'a skilled, unproblematic exercise in precision, made increasingly accurate by modern technological advances' (1997, 17). For Enlightenment and post-Enlightenment cartographers, the emphasis on accuracy became inextricable from objectivity: the map became a symbol for the Enlightenment's rejection of subjectivity and its promotion of purely rational values. As Black argues, this in itself is an ideological position: because a map can never be entirely representationally accurate or objective – a fact to which the stories of Carroll and Borges attest – the narrative of the neutral and truthful map works to make invisible the subjectivities that are contained within it. Wood argues that 'the map is about the world in a way that reveals, not the world – or not just the world – but also (and sometimes especially) the agency of the mapper. This is, maps, all maps, inevitably, unavoidably, necessarily embody their author's prejudices, biases and partialities ...' (1992,

24). By disguising these prejudices and biases as objective fact, the map becomes a means of creating a dominant power relationship between the map-maker and those whose interests they embody – be that a nation-state, a cultural group or the human species at large – and that aspect of the world which the map is purporting to represent. Wood and Fels set out to prove the inherent ideological nature of the 'objective' post-Enlightenment map by examining how these maps depict the natural world, an entity that above all else ought to be free of human ideological perspective, in order to demonstrate how even in this case, the map 'creates ideology, transforms the world into ideology, and by printing the world on paper constructs the ideological. It doesn't matter what has the map's attention. Whatever its subject is will be turned into something it isn't and in the process, inescapable, unavoidably, made ideological' (2008, 7). In the case of the natural world, Wood and Fels argue, the map, its pretence at objectivity and its actual subjectivity become a way of claiming authority over territories and land, a project that encapsulates the hierarchy between human and nature that forms the subject of the following chapters, and that found new forms of expression in Enlightenment thought.

Not only does modern cartography remain bound up in ideology as a value-laden image, but the conditions that catalysed its development are strikingly ideological. In Britain, the reconceptualization of cartography was prompted by the commissioning of the Ordnance Survey in the eighteenth century. Although by Tolkien's time, the Ordnance Survey was largely an innocuous object used by the public for outdoor leisure activities, its original purpose was military. After the Jacobite uprising and the defeat of the Scottish rebels in the Battle of Culloden in 1746, English royalist troops struggled to navigate the difficult Highlands territory to round up the remaining rebels. Hewitt comments on the inadequacy of military cartography in this period: throughout the rebellion, the English troops had little to no information about the geography of the Highlands, which was only 'exacerbated by inadequate maps of the region …' (2010, xviii). The benefits of an accurate, comprehensive map of Britain were quickly realized, and in 1747, David Watson, a Quartermaster General in the English army, proposed a military survey of Scotland (2010, 17). By 1752, the whole of the Scottish Highlands had been mapped, and by 1755, the Lowlands were also finished. Impressed with the results of these surveys, and recognizing their applicability in military matters, a survey of England and Wales was also commissioned, out of which eventually grew the Ordnance Survey of Great Britain (2010, 44). In 1784, the primary triangulation of Britain was carried out, and in 1801, the first Ordnance Survey map was released to the public, depicting areas of Kent (2010,

163). In the following several decades, the rest of Great Britain was also mapped out and published, and in 1870, the Ordnance Survey was completed, at a scale of 1 inch to 1 mile (2010, 305). Throughout its initial production, however, the Ordnance Survey was already being improved; by 1840, a new series of larger-scale, 6 inches to one mile maps had been commissioned (2010, 295), and in 1856 a series of 1:2,500 scale maps (roughly 25 inches to a mile) were produced, which represented the landscape in much greater detail (2010, 300).

The increasing scale of the Ordnance Survey maps in the nineteenth century shifted how they were read and used. The original, relatively small-scale 1 inch to 1 mile map served its initial purpose – to gain intelligence about an area for military purposes through a reliable and accurate representation of the landscape – yet it was not detailed enough to act as an informational or navigational resource for the public, who would have more use for a map of their immediate area, rather than a general map of a region. Hewitt explains that although wealthy landowners did buy the first Ordnance Survey maps, they did not serve any practical purpose but were rather used as 'rhetorical images of power and ownership' (2010, 167). The introduction of a larger-scale map was rooted in the commercial rather than military or governmental demands of the nineteenth century. Widespread urban growth, industrial activity both in the towns and in the coalfields of the Midlands and northern England, and the fast-developing road and railway network throughout Britain were transforming the way that land was being used and negotiated (Delano Smith and Kain 1999, vol. 2: 221). At the same time, mass tourism was becoming a more and more popular activity in Britain made possible by the railway: Hartmut Berghoff and Barbara Korte argue that the natural landscape, both in Britain and abroad, became a site of touristic interest in the nineteenth century, explaining that 'Romanticism essentially contributed to the redefinition of nature that was needed to set up and establish tourist destinations …' (Berghoff and Korte 2002, 5). Urbanization, industrialization, and new transport networks that enabled mass tourism all brought about a demand for large-scale maps that could be printed and bought cheaply, were widely available, and which could be used for both administrative and personal activities.

Although the touristic purposes of the Ordnance Survey map are less overtly ideological than its original military function, the fundamental intent behind both – to reproduce the landscape as accurately as possible on paper in order to enable human navigation and control – embodies the ideals of Enlightenment cartography. At every stage of the Ordnance Survey's development, its need for accuracy was reiterated: towards the end of the nineteenth century, the

Board of Agriculture, under whose jurisdiction the Ordnance Survey operated, appointed a Departmental Committee to investigate the survey, with the result that regular revisions were authorized, so that no 1-inch map would ever be more than fifteen years out of date, and no 6-inch or 25-inch map would be more than twenty years out of date (Oliver 2005, 18). An instructional manual for the Survey dating from 1952 meanwhile states that '[t]he object of the large scale survey is to produce a plan on which no measurable inaccuracies shall appear …' (Oliver 2005, 11).

This objective is reflected in the abstract visuals of the maps: abandoning the illustrative tendencies of pre-Enlightenment cartography, iconography is intended as a vehicle for the efficient conveyance of topographical and environmental data. The first Ordnance Survey map of Kent from 1801 did not provide a legend for the symbols on the map, but the symbols were largely self-explanatory: clusters of small trees indicated forest areas, while trees arranged neatly in rows signified an orchard. Buildings were depicted as small shaded blocks, and churches as a small cross. Notably, relief was shown by hachures to indicate the gradient of a hill, which eventually shifted to contour lines in later versions at the end of the nineteenth century (Hewitt 2010, 164–65). Altogether, although the symbols were to an extent pictorial, they nevertheless functioned as a series of abstract signifiers characterized by uniformity, synecdoche and geometry: the trees were all a uniform size and shape, churches weren't depicted as a building but as an icon, and hilly areas weren't shown from a front-facing angle typical of medieval maps, but from above using hachures. Large-scale maps meanwhile use the same symbols as the smaller-scale maps, yet by dint of their scale these symbols are further spread out across the page, resulting in large areas of blank space representing fields, or the gradually inclining areas between contour lines. The appearance of these maps, despite being theoretically more detailed, is therefore at first glance sparser than the smaller-scale maps, particularly in rural areas. This purposeful use of blank space was a relatively new concept. Whereas in the medieval period, blank spaces on a map were encoded with an underlying significance, indicating the unknown, the unmappable, and the dangerous, in the large-scale Ordnance Survey maps, large blank spaces were reclaimed as a representational tool, with white expanses on the map giving a sense of the largeness of the landscape. Each of these symbols, from the uniform trees to the contour lines and blank spaces, not only embeds the topographical and environmental characteristics of what it represents onto the map but also encodes the map's purpose of precise, ordered and neutral reproduction.

Despite this constant emphasis on accuracy, it was and is necessarily impossible to depict every element of the landscape with the same level of detail, so that each iteration of the Ordnance Survey shows a different prioritization of certain aspects of the landscape. After the First World War, for example, a simplified map was produced; named the Popular Edition, it introduced a new and highly detailed road classification, but relied solely on contours to depict relief, with the result that landform information was less detailed, while the cultural and commercial content of the map was emphasized instead (Delano Smith and Kain 1999, vol. 2: 224). The variations of scale also show a different prioritization: the small-scale maps give a better sense of the overall landscape, and in particular offer a better depiction of relief as the contour lines are shown close enough together to give sense of the sharpness of the gradient, while large-scale maps are more appropriate for practical or navigational purposes in an immediate area and prioritize a detailed depiction of roads and buildings. These variations in the maps' purpose demonstrate the curatorial power of the map-maker in determining what is depicted on the map and what it can be used for. Wood points to this selectivity as the primary characteristic of map-making, and as what indeed enables the map to work: the map curates its content towards a particular interest, which is then 'embodied in the map as presences and absences' (1992, 1). These presences and absences give the map a focus, and allow it to function as a representation of the world that fulfils a particular role, be that navigational, imaginative or political; yet they inherently render the map subjective and, by definition, incomplete. This is embodied in the Ordnance Survey maps; while they are accurate representations of the landscape, in that there is faithful representation of distance, direction and scale, these maps nevertheless reveal the falsity of 'objective' cartography's dream of completeness, both in how they were used historically and in how they are constructed. By selecting what does and does not appear on each particular map while simultaneously affecting comprehensive representation, these maps remain socially and politically constructed texts that, as Wood and Fels argue, are 'inescapably, unavoidably made ideological' (2008, 7).

The Ordnance Survey revisited its original function during the First World War, becoming a major player in helping to form military strategy, reinforcing Wood and Fels' argument about the inescapable ideological functions of the supposedly politically neutral map. The Ordnance Survey produced topographical maps of the Front from mid-1915, helped from late 1917 by the Ordnance Survey Overseas Branch, so that during the First World War, as Black points out, the 'overwhelming majority' of maps produced by the Ordnance Survey were representations of battlefields (2016, 38). The Royal Geographic Society, at the time an eminent centre of scientific research, also contributed to the war efforts; in 1914, the Society was placed at the disposal of the Geographical Section of the General Staff (GSGS), a department of military intelligence and one of the precursors of MI5 and MI6 concerned with military map-making, map collection and topography (Heffernan 1996, 507–08). The disciplines of cartography and geography were thus firmly bound up with military activity, so that the maps' accuracy became used in the service of ideology.

Indeed, more than any other war before it, the First World War realized the full potential of maps as a military tool. Lieutenant Colonel E. M. Jack, the officer and engineer in charge of all British surveyors and map-makers on the Front, famously declared that 'a map is a weapon' (Chasseaud 2013, 10), and the realities of cartography during the First World War validated his claim. When the British Expeditionary Force first arrived in France in 1914, there was only one officer and one clerk in charge of map-making, and the maps were largely unreliable. By 1918, the section of the army preoccupied with map-making had risen to approximately 5,000 men, who produced over 35 million map sheets in the total period of the war (Black 1997, 154). Moreover, it was not only the British forces that recognized cartography's military potential: in 1914, the German army mostly had to make do with simple, inadequate sketch maps and often found themselves entirely lost in the French countryside, before similarly realizing the need for modern, detailed cartography (Espenhorst 2016, 83). Each of the major players in the war eventually had an official mapping organization: in Paris it was the Service Géographique de l'Armée, in Berlin the Königlich Preußische Landesaufnahme, in Vienna the k.u.k. Militärgeographische Institut, and in St Petersburg the Military Topographical Section of the General Staff (Chasseaud 2013, 9). In the four years of the war, mapping became an indispensable part of an army's tools.

As with the original mapping of the Highlands, the map's military purpose was inextricably linked to an emphasis on accurate representation. This was partly informed by the overall cultural shift towards an exact cartography, but the particular requirements of the First World War mapping – namely, accuracy of relief and gradient – were also due to developments in weapons technology that changed the nature of warfare dramatically. As opposed to having a direct line of sight between a weapon and its target, much of the fighting in the First World War battles was carried out through indirect fire, where the distance and the angle of an unseen target would be calculated in order to determine the trajectory of the shell. Accurate, large-scale maps therefore became indispensable for facilitating this type of warfare, and for allowing the artillery to find their mark (Black 1997, 153–54). In particular, topographical maps that showed the relief of the landscape, typically through contour lines, were essential in order to calculate the elevation required for the weapon to fire. Moreover, the new technique of trench warfare meant that any military intelligence that could be gathered about an enemy's trenches would need to be graphically visualized as the trenches themselves could not be seen, again necessitating accurate maps. The scattered nature of the war's battlefields also increased the need for maps: the French originally concentrated their mapping on major fortified positions near the German border where previous battles had been fought, only for a new, mobile warfare to take place, and these maps be rendered useless (Black 2016, 34). A comprehensive cartography was therefore needed that was accurate enough to accommodate unforeseen sites of battle and anticipate hidden military targets.

In order to accommodate these new demands of cartography, new stylistic features were used. Specialized symbols unique to warfare were developed: a map legend from a 1917 trench map (NLS Sheet 28.NE3) distinguishes between standing and ruined houses, used and disused trenches, and different types of ammunition and weapon stations, revealing the importance of transmitting specific information as efficiently as possible. Contour lines were also used prolifically in order to depict the elevation of an area. These lines were used in combination with other elements to depict relief in as great a detail as possible: in particular, the British forces commissioned maps that employed colour, superseding the monochrome maps of previous battles, to further emphasize the relief of the landscape (Black 2016, 39). Colour was also very important for distinguishing between ally and enemy trenches: numerous trench maps draw the trenches of opposing sides in different colours, such as a 1918 map (NLS 51B.NW) which shows the British trenches in red, while the German trenches are in blue.

All of these stylistic features were employed to make the maps as accurate and readable as possible. Yet what is striking about military maps is that despite this semblance of total factuality, they continue to be inextricably enmeshed with ideologies, subjectivities and politics. Michel Heffernan comments on the role of the Royal Geographical Society in the First World War, arguing that it 'illustrates how geographical knowledge and expertise can become implicated in broader political and ideological conflicts, and how ostensibly universal, "scientific" objectives can easily become fused with the narrow, strategic objectives of the nation-state' (Heffernan 1996, 522). Warfare, seen both in the First World War and in the conflict between the English and Scottish that catalysed the production of the Ordnance Survey, reveals how easily cartography could be appropriated for use as a tool in discourses of power. Unlike medieval maps, which prioritized the visual encoding of religious ideology above exact representation, modern military cartography focused on accuracy based on a particular subject position in the service of a political agenda. This subjectivity had practical and visible results in the First World War cartography: Peter Chasseaud observes that British maps created between 1915 and 1917 would only show German trenches for security reasons and that British trenches were depicted on 'secret' editions of the maps available only to officers, rather than front-line troops (2013, 13–14). The realities of the map – what it depicts and how it depicts it – were therefore affected by the map's potential audience and the political circumstances it was created in. Military cartography is therefore neither total nor complete in its representation of the world; it reflects the cartographer's political allegiance or the sensitivity of the information it needs to relay. A map that Tolkien himself used during the Battle of the Somme in 1916 illustrates this. The high casualty rate at the Battle of the Somme can be partly attributed to misleading information given to the troops regarding the strength of German barbed wire and the state of their defences (Bodleian Library 1992, 31). Tolkien's trench map depicts the German trenches and areas of barbed wire, with annotations such as 'gaps in wire every 30 or 40 yards' and 'wire here thin'. These annotations were probably made based on information from captured German soldiers, with a 'consequently dubious level of accuracy' (Bodleian Library 1992, 31). This map raises several points on the question of cartographic subjectivity. First, the dependence on captured soldiers for intelligence reveals that even with high-level surveillance technology and mapping techniques, cartographers in the First World War did not derive all their data using purely scientific means, but also collected them from unreliable sources which could affect the accuracy of the map. Second, Tolkien's own annotations on the map show the ways in which a map can be altered and reframed. The notes highlight certain areas as being of high importance or interest, and the elements of the landscape are interpreted for his or the army's immediate agenda. The map becomes a palimpsest with layers of meaning inscribed by each creator and user. It cannot represent an objective truth, but rather remains what Harley terms a value-laden image, one that inescapably constructs and is constructed by its sociopolitical context.

Tolkien's Cartography

Caroll and Borges were correct: there is no such thing as a neutral map. As we have seen, both medieval and modern cartography embed the conditions of their production – from their sociopolitical environment to culturally dominant ideologies – within their image. The rest of this chapter examines the ways in which Tolkien draws from these practices in order to create a corpus of fictional cartography that fits within a similar tradition of ideologically informed mapping. Tolkien's cartography extends beyond mere paratextual and illustrative purpose, but rather incorporates ideological forms of map-making in order to place his maps in conversation with the political narratives of the legendarium, as well as to position maps more broadly as an innately political medium.

It is important to note at this point, however, that Tolkien's cartography is also a literary project that was informed and confined by practical demands. His maps were shaped not only by historical frameworks but also by financial and material limitations set by his publishers, and by the efforts of his son Christopher Tolkien, who collaborated with his father and redrew the final published maps in *The Lord of the Rings*, as well as the map of Beleriand found in *The Silmarillion* and the map of Númenor found in *Unfinished Tales*. The paratextual purpose of these maps cannot be entirely ignored; his cartography was both stylistically and conceptually shaped by other needs that Tolkien was attempting to fulfil, such as drawing the maps at an appropriate scale for the reader to follow along, giving an adequate number of place names and having a limited number of colours. Strikingly, however, Tolkien often narrativized the elements that derived from material requirements, integrating external, paratextual factors within the conceit of a fictional, internally consistent cartography that further illuminates his awareness of the ways in which maps inescapably convey political meaning. This section will approach Tolkien's maps mindful of their paratextual context, while simultaneously examining how this is drawn into a broader framework of politicized cartography informed by both medieval and modern traditions.

Map I: I Vene Kemen

Chronologically, the first conceptually complete map – that is to say, a developed rather than rough sketch map – in Tolkien's corpus is I Vene Kemen, reproduced in *The Book of Lost Tales I* (1983). Although it is not dated, it likely originates from between 1916 and 1919, from the same period when the majority of the tales collected in the volume were written. I Vene Kemen, which translates to

'The Vessel of the Earth' according to the Gnomish Lexicon, or 'The Shape of the Earth' according to the Qenya Lexicon, portrays various lands of Arda discussed in *The Book of Lost Tales* – including Valinor and Tol Eressëa – as well as the surrounding seas and atmosphere, depicted as a cutaway drawing of a large, Viking-like ship. The main body of the ship is Vai or Neni Erùmenor, or the Outermost Waters, in which lies Ulmonan, the halls of Ulmo, the Vala of the sea and Uin, the Great Whale, who carried the island of Tol Eressëa across the sea. Above this are the lands I Noro Landa (the Great Lands), Valinor, from which emerges the peak of Taniquetil, Tol Eressëa and I Tolli Kuruvar (The Magic Isles), as well as the Two Trees to the extreme west of the map. Surrounding these lands is Ô (The Sea) and Haloisi Velike (The Great Sea). The mast of the ship rises from the highest point of the Great Lands, from which flies a sail, featuring a drawing of the Sun (Ûr), the Moon (Sil), and Luvier (Clouds). Next to the sail are three layers of clouds, labelled Vaitya, Ilwe and Vilna; these are the different layers of the Earth's air which encompass the world (J. R. R. Tolkien 1983, 85–86).

I Vene Kemen is certainly the most unique and least realistic of Tolkien's maps. As Fimi argues, it very clearly taps into the distinctly mythic tone of *The Book of Lost Tales* (2009, 124), and was an experimental idea which was quickly abandoned. No trace of the Earth as ship remains in any of Tolkien's future mythology, although many of the key ideas of the map – the encompassing oceans and air, for example – are maintained and reconceptualized in his later mythology. The motives for Tolkien's initial decision to present the world as a ship remains unclear. Christopher Tolkien links it to a speech by Ulmo, where he addresses the Valar, 'O Valar, ye know not all wonders, and many secret things are beneath the Earth's dark keel, even where I have my mighty halls of Olmonan, that ye have never dreamed on ...' (J. R. R. Tolkien 1983, 86); potentially, I Vene Kemen was an attempt by Tolkien to visualize this idea. However, although the reasoning behind the ship remains uncertain, the structuring of the world in such a highly conceptual form is definitively influenced by medieval ideas about the world and how it might be represented.

Fimi argues that Tolkien was drawing on ideas from North European texts of the Middle Ages: in the Old Norse Prose Edda, 'four dwarfs support the sky, while the sky itself is described as the dome of a giant's skull set up over a flat earth ...' (2009, 124); in English Christian tradition, meanwhile, the world was conceptualized as the body of Christ, such as in the Ebstorf map, which features Jesus' head at the top, his feet at the bottom and his hands either side. There exists a clear medieval tradition of conceptualizing the world through a cultural – be that mythological or theological – lens, a practice that the I Vene

Kemen map emulates by eschewing a realistic or scientific world model; much as with medieval mappae mundi, I Vene Kemen is not intended to make the world more navigable or comprehensible, but rather to reflect cultural understandings of place and space. In Tolkien's mythology at this point, the Earth is a flat disk, floating on a large Enfolding Ocean which is 'more like to sea below the Earth and more like to air above the Earth' (J. R. R. Tolkien 1986, 236): yet whether the I Vene Kemen map is specifically visualizing Ulmo's comment or not, its form speaks to his characterization of the world as complex and ontologically conceptual rather than fixed. At this stage in Tolkien's mythology, when the world is still relatively new and under both authorial and diegetic construction, its portrayal as a symbol rather than a topographical landmass emphasizes both a world-building and interior culture more interested in their spiritual, existential and cultural place in the world, rather than its geographical quality.

Map II: The 'Ambarkanta' Diagrams and Maps

It is difficult to place the next map in Tolkien's corpus chronologically, as certain posthumously published works remain undated, but it is likely that the 'Ambarkanta' diagrams and maps were drawn after I Vene Kemen but before the maps in *The Hobbit* and *The Lord of the Rings*.[8] The 'Ambarkanta' is a short work collected in *The Shaping of Middle-earth* (1986), which describes the cosmological and geological properties of Arda and its formation at the beginning of the First Age, and is accompanied by three diagrams and two maps. Like I Vene Kemen, the 'Ambarkanta' diagrams and maps were not prepared for publication in the same way as those of his novels, so they too do not have the same paratextual concerns as later maps will. Also similarly to I Vene Kemen, the 'Ambarkanta' diagrams and maps are small-scale world maps that illustrate the mythological stage of Tolkien's world-building, containing many of the features of the previous map, but abandoning the ship form in favour of a more familiar, globe-like depiction of the world.

The first diagram, labelled Diagram I, is a straightforward visualization of the description in the text of Ilu (the world, or more accurately 'everything') before the Changing of the World, when it was transformed from a flat disk into a globe. The world here is depicted from West to East, with very little aesthetic

---

[8] Although there is no date given for the 'Ambarkanta', Christopher dates the 'Quenta' that is collected in the same volume to roughly 1930 and explains that the 'Ambarkanta' was written later, 'perhaps by several years' (1986, 235).

embellishment. The diagram very accurately conveys the textual description: the layers of the Enfolding Ocean and Air are all in the correct position and labelled, and their relative thickness is even maintained: the text explains that 'Ilmen lies above Vista, and is not great in depth, but is deepest in the West and East, and least in the North and South' (J. R. R. Tolkien 1986, 236), and the diagram depicts Ilmen (the sky) thinner at the top and thicker at the edges. There are some visual codes for illustrating the different areas' geographical properties: five small crosses depict stars in Ilmen, where 'the courses of the stars' were set (1986, 236), two clouds are set in Fanyamar, or Cloudhome, and Ambar, or the earth, has vertical, fissure-like lines to distinguish it from the air and water. The second diagram, labelled Diagram II, is very similar to the first, albeit simplified, with none of the already few illustrations that accompanied the first diagram. It also shows a shift in perspective, with the map now oriented towards the north. The third diagram, labelled Diagram III, is very similar in style to the first two, but shows Ilu after the Changing of the World, when the Earth has been made round. The diagram now resembles a series of concentric circles, with Ambar – now also a circle – at the centre, surrounded by the same layers of water and air: Vista, Ilmen and Vaiya. The Straight Path, created after the destruction of Númenor, now passes over Ambar and through Ilmen. This diagram has absolutely no illustrative features: it serves entirely as an almost technical depiction of the structure of the world.

The first 'Ambarkanta' map, labelled Map IV, retains the diagrams' depiction of the world's atmosphere and Enfolding Ocean, although the focus has shifted to representing terrestrial features: Valinor is in the western corner, and separated from Middle-earth by a sea depicted with numerous, closely set, parallel lines. Middle-earth is then separated from the Lands of the Sun in the east by the East Sea, similarly illustrated. Geographical features are also marked out using pictorial symbols – lines of upside-down Vs represent mountain ranges, and the Sea of Helkar and the Sea of Ringil are depicted using contour-like lines – making this map the first of the early maps to use such techniques for representing geographical details. The final 'Ambarkanta' map, labelled Map V, bears a small note at the top, 'After the War of the Gods', placing the world it depicts after the imprisonment of Melkor by the other Valar. This map is less neat than the previous one, but retains many of its characteristics, including the depiction of the Vaiya, the lines in the sea, and the rough yet pictorial mountain peaks. This map also makes an attempt at depicting a bird's-eye view of coastal outlines and land masses, rather than the straight blocks of land seen in the previous diagrams.

The 'Ambarkanta' diagrams and maps continue the mythology of a flat Earth – eventually made round – that was depicted in I Vene Kemen, yet their layout is far less allegorical. Instead, the five figures emulate the appearance of mappae mundi: the three 'Ambarkanta' diagrams are reminiscent of the Macrobian, or zonal, mappae mundi, which split the world up into five climactic zones; in the same way, the 'Ambarkanta' diagrams attempt to map out the non-terrestrial, atmospheric conditions of the world and its surroundings by demarcating the atmospheric zones of Arda. The 'Ambarkanta' maps, meanwhile, are more akin to the Isidorian, or T-O, maps, such as the Hereford or Ebstorf mappae mundi. Drawing on the Macrobian model, Isidorian maps still gesture towards non-terrestrial features – Fisher points out that the Hereford map is surrounded by a layer of water, much as Arda is surrounded by Vaiya, the Enfolding Ocean (Fisher 2010, 13) – but also detail the structure of the surface, sketching out land, sea and geographical features. In particular, Map IV's layout of the land recalls that of the Isidorian mappae mundi: it is not geographically accurate and does not attempt any neat outline; rather, it splits the world longitudinally into zones of land and sea, thereby showing their position in relation to each other, while not attempting any fidelity of surface area or shape. Map V, meanwhile, draws on even later map-making techniques, such as the regional maps and even the portolan charts, in its depiction of land mass: while it does not approach the accuracy of the latter, it does attempt to map out the shape of the continents in an increasingly precise way.

Despite their visual similarities, however, the 'Ambarkanta' diagrams and maps do not strictly reproduce every characteristic of the mappae mundi. Tracing the evolution of Tolkien's cartography between I Vene Kemen and the 'Ambarkanta' diagrams and maps, it is clear that the trajectory is towards a more modern, representational cartography, the conceptual nature of I Vene Kemen bringing into relief the methods of medieval world mapping that the 'Ambarkanta' charts reject. Although the 'Ambarkanta' maps closely resemble the mappae mundi stylistically through their rounded shape and zonal structure, they lack the theological underpinnings that characterized the mappae mundi, whether through their allegorical layout or pictorial symbolism. Given the geological nature of the text they accompany, certain diagrams and maps take on a more scientifically representational purpose: as discussed above, although Map IV does not make any gesture to accuracy or scale, Map V begins to chart outlines of land masses and topographical details such as mountain ranges in the spaces where they should appear. Diagram III, meanwhile, depicts the Earth after it was globed and thus features a cutaway perspective that, combined with the maps' lack of pictorial symbolism, is more reminiscent of modern geological charts. The 'Ambarkanta' diagrams and maps become emblematic of the pseudomedieval nature of Tolkien's cartography, as put forward by Ekman; there is no denying the overt medieval influence on the stylization of these maps, but closer inspection reveals this influence to be mostly aesthetic rather than conceptual in character, seen in the charts' lack of theological or mythological symbolism, particularly when compared with I Vene Kemen. Meanwhile, an emerging late-medieval and modern influence can be traced within the maps through their engagement with more accurate representation. It is notable that the 'Ambarkanta' diagrams and maps are concerned with materially reproducing the shifts in geology and geography that occur at this stage in Tolkien's world-building; given the particularly unpredictable state of the world's make-up in this period, there is a diegetic need to map control over these changes in the manner that modern cartography allows.[9]

Map III: Thror's Map

The first of Tolkien's published maps in this corpus, Thror's Map was one of five that Tolkien sent to his publishers to be included in *The Hobbit*. The other maps were the Wilderland, which also made it into the final published version, a map of the Misty Mountains and the Great River, one of the Lonely Mountain and its surroundings, and one of the Long Lake (Hammond and Scull 2011, 11). Tolkien eventually decided that the last three maps were 'not wanted' (2006, 14); Thror's Map, however, was a vital part of Tolkien's narrative and cartographic construction of *The Hobbit* from the very beginning. The first attempt at Thror's Map appears in the original manuscript of *The Hobbit*, when it was known as Fimbulfambi's map, Thror's original name. John D. Rateliff notes that although this map differs 'in significant details from the final version, it is remarkable how many permanent elements were already present and persisted from this first hasty sketch …' (2011, 18). The mountain is marked with six spurs outlined in hachures, with the River Running leading away to the right of the mountain. Lake-town is located on another branch of the river lower down, and the ruins of Dale are also marked. A sinister hand points to the mountain, although in this version it is more detailed and individualized, with long, pointed nails and shading around the bent fingers and knuckles. Runes below the hand explain 'FANG THE SECRET PASSAGE OF THE DWARVES' (Hammond and Scull

---

[9] See Chapter 3 of this book.

2011, 49), and just below in English are the inscriptions which would eventually become the runes and moon letters in the final map.

A later version of the map bears far more resemblance to the final product. Rather than a sketch made at the edge of a page of writing, this map is a more purposeful drawing, taking up an entire page. Unlike the first draft, which may have been used as a working map for Tolkien's own planning, this map is clearly intended as a draft of artwork for the final published book: its overall appearance is neater, it is on its own page, and most importantly, the inscription in the bottom-left corner, which reads 'Thror's Map. Copied by B. Baggins. For moon runes hold up to the light', indicates that the map is intended for external readers, Tolkien having planned at this stage to have the moon letters printed faintly on the back of the map, so that they would be seen when held up to the light (2011, 49). This inscription also introduces the notion of the printed map in *The Hobbit* replicating the map described in the text by framing it as an artefact reproduced from Bilbo's collections. Stylistically, the map has also developed greatly; the mountain is still very similar, with hachures used to depict the six spurs, but the now iconic dragon is marked in red ink on its peak. The river has been unified and is now one branch running southward, with Dale and the Long Lake still marked. The runes are written more neatly around a simplified hand. Perhaps the most striking difference here is the emphasis on historicization and exposition that is characteristic of the final map: arrows indicate the direction of Mirkwood, the Grey Mountains, Withered Heath, and the Iron Hills of Dain off the sides of the map, and two labels proclaim, 'here of old was the land of Thrain King under the Mountain' and 'here is the Desolation of Smaug', written in stylized, archaic script.

Publishing restrictions, however, somewhat altered the appearance of the final map. Tolkien had hoped that Thror's Map would be 'tipped in (folded) in Chapter 1, opposite the first mention of it: "a piece of parchment rather like a map"' (2006, 15). However, his publishers decided instead to print both Thror's Map and the map of the Wilderland as endpapers, which meant the moon letters could no longer be printed on the reverse. Instead, the letters were printed on the front in a hollow font, in order to stress their ephemeral nature. For the map to fit better to the size and layout of an endpaper, which has a landscape rather than portrait orientation, Tolkien rotated his map ninety degrees, which meant that east now faced the top. Stylistically speaking, this final map is both textually and pictorially far more complex than any of the previous iterations. The script has become even more elaborate, with many of the capital letters featuring double minims. The map contains even more written information, at times non-geographical: the reader is informed that Mirkwood contains spiders; Lake-town is also referred to as Esgaroth and it is specified that Men dwell there; the Withered Heath is identified as where the dragons came from; and Girion's location in Dale is labelled. Pictorially, there is also much to note: as well as the illustration of Smaug now flying above the mountain, there is also another dragon next to the label about the 'Great Worms'; the mountain itself is now drawn from a face-on perspective, rather than with hachures; drawings of withered tree stumps visually reinforce the Desolation of Smaug, while spiders' webs and a small spider complement the warning of spiders near Mirkwood.

Thror's Map represents a return to medieval cartography, both aesthetically and conceptually, aligning particularly with the medieval tradition of itinerary maps. To begin with, although the map covers a larger area than many medieval regional maps, encompassing numerous large geographical features such as a mountain, two large stretches of river, and a wasteland, its area of focus is nevertheless limited to what is relevant to the dwarves' journey, depicting just the Lonely Mountain and its environs, with surrounding areas indicated off the edge of the map. This narrow focus is further emphasized by the lack of detail on the map: aside from the few geographic features discussed above, the map relies on its runes to convey information. Both textually and pictorially, Thror's Map is intended to aid the reader in reaching the Lonely Mountain and locating the hidden door, a purpose that is further established by the way the map is diegetically used within the text by the characters. Although Thror's Map lacks the intricate road networks that characterize medieval itinerary maps such as the Matthew Paris maps of Britain or the Gough Map, its combination of step-by-step textual instructions, illustration of geographical features, and focus on a specific destination (in this case, the door of the Lonely Mountain) make it unequivocally a navigational and specifically an itinerary map. The map's itinerary construction also fulfils a generic function: as *The Hobbit* is fundamentally a quest narrative, the itinerary map's allowance for navigation within a particular route makes it the ideal paratext for the genre.

This medieval conceptualization of navigational function is then reflected in the map's visual and structural framework, which also largely tends to the medieval. The medieval aesthetic is primarily achieved through the use of illustrations rather than abstract symbols to convey information about the landscape: Matthew Paris' itinerary map from London to Palestine represents urban areas through individual, face-on illustrations of buildings, while the Gough Map represents towns as a combination of small houses and larger buildings, bodies of water such as Loch Tay in Scotland as green circles detailed with wavy lines, and mountains to the north of the loch as a self-contained range of five mounds. Thror's Map is a prime example of this tendency to the pictorial, seen in its illustrations of the Lonely Mountain, the withered trees, the dragons, and the spiders' webs near Mirkwood; the development of the Lonely Mountain from initial draft to the final map particularly highlights the use of pictorial symbols as a visual choice. Crucially, however, these pictorial markers fulfil functions other than the aesthetic, conveying ideological information in the same way that mappae mundi did. The dragon, initially not depicted on the first draft map, takes up increasing space with each successive draft, and appears in its largest form in red ink on the complete map; its vivid presence signals the domination of the mountain by the dragon, and centres the map's purpose on its removal. The violence that the dragon has wreaked is emphasized by the burnt trees throughout the Desolation of Smaug; again, these were added only in the third sketch but their presence on the map reinforces the shift in political control over the Mountain and the broader, in this case environmental, effects that this has had, incentivizing the dwarves to follow the map's itinerary and fulfil the quest. Other illustrations mark out the ways in which the land is occupied and controlled, and act as warnings for the reader: the second dragon on the map points to the Grey Mountains where other dragons might be, while the spider's web in the bottom corner of the map highlights the potentially lethal consequences of entering Mirkwood. The hand on the edge of the map pointing towards the runes, meanwhile, echoes the manicules found in medieval manuscripts, intended to draw attention to particular sections of the text: its presence on Thror's Map demonstrates how the map is fundamentally designed to direct the reader's understanding of it. Rather than engaging with theological perspectives, Thror's Map embeds political narratives through its iconography, thus maintaining medieval cartography's inherently ideological visual language.

Structurally, Thror's Map is also influenced primarily by mappae mundi and their tendency to inscribe meaning into their layout, through their orientation towards the East, as seen in the Ebstorf map and the Hereford map, and in their relegation of unknown areas to the edges of the map. As discussed above, Thror's Map's orientation towards the East was purely a result of publishing restrictions based on financial costs; however, Tolkien integrated this change within his mythology, explaining in the preface to the 1966 edition of *The Hobbit* that this was 'usual in dwarf maps ...' (Hammond and Scull 2011, 55). This retrospective rewriting demonstrates Tolkien's awareness of the ways in which maps structurally convey signification; although Thror's Map was not specifically influenced by medieval cartography's orientation towards the east, it nevertheless acknowledges the ways in which cartography structures and signifies space. It is most deliberately medieval in its encoding of danger at the fringes of the map, embodying Camille's argument about the medieval politics of spatial representation. On the map, arrows point down towards Mirkwood and towards the Grey Mountains, warning 'West lies Mirkwood the Great there are Spiders' and 'whence came the Great Worms', mimicking the symbolic construction of medieval cartography. Much as the Ebstorf mappa mundi populated its edges with the unknown and monstrous, delineating the limits of civilization using the limits of the page, Thror's Map similarly demarcates its maker's boundaries of knowledge, reinforcing the map's subject – the dwarves' ancestral home – as the centre of their narrative. Moreover, although both Mirkwood and the Withered Heath would always have been located off the map's edges, each iteration of the map makes both this and their unsafe quality more explicit: the first draft mentions nothing; the second draft has arrows labelled with place names pointing off the edge of the map; while the final draft expands on these place names to include the dangerous creatures (spiders and Great Worms) which reside there. By emulating this tradition, Tolkien creates a tension in his sub-creation between known and unknown spaces similar to that in the medieval world, which allows for an exploration of the pull between home and adventure which Bilbo experiences.

Thror's Map is therefore highly influenced by medieval cartographic practices, and in particular by these practices' ideological underpinnings. This is not to say that it is a perfect simulacrum of a medieval map, however. Its pseudomedieval quality can be seen in the compass rose in the top right-hand corner – a feature not found on medieval terrestrial maps – and in its combination of practices from distinct types of medieval cartography, thereby creating a generic form of medievalesque cartography rather than authentically replicating a single map type. Overall, however, it is, alongside I Vene Kemen, the most medieval of Tolkien's cartographic output. As Tolkien embarked on the more complex narrative of *The Lord of the Rings*, his maps became increasingly more intricate in the variety of sources that they drew from, as is seen in the Middle-earth map and the map of Rohan, Gondor and Mordor, both of which embody a more complex pseudomedievalism.

Map IV: The Middle-earth Map

In *Unfinished Tales*, Christopher Tolkien refers to his father's draft maps of Middle-earth as 'sketch-maps', a phrase which he later corrects in *The Treason of Isengard* (1989): 'this was an ill-chosen word, and in respect of the First Map a serious misnomer. All parts of the First Map were made with great care and delicacy until a late stage of correction, and it has an exceedingly "Elvish" and archaic air ...' (299). Although Tolkien did make numerous rough sketches of various areas of Middle-earth throughout his writing process,[10] the drafts of the small-scale, general map of Middle-earth are highly detailed and meticulously planned. The earliest of these, known as the First Map, is described by Christopher as 'a strange, battered, fascinating, extremely complicated and highly characteristic document' (1989, 295). It is composed of several sheets of paper glued together, with redrawn sections of the map pasted over previous sections. This map is probably one of the best examples of Tolkien's cartography and his narrative developing simultaneously and symbiotically. In a 1944 letter to Christopher, Tolkien explains that he has solved certain problems with the narrative's chronology by 'small map alterations, and by inserting an extra day's Entmoot ...' (2006, 97). Many of the alterations are toponymical in nature: just to the north of Rivendell is an area marked Entish Land next to a note that specifies 'alter Entish Lands to ... Ettenmoor', which Christopher identifies as the first use of the name in the mythology; the River Iren was eventually changed to Isen, and Andon to Anduin, all of which are nomenclatures which are eventually incorporated into the narrative (1989, 306). Elsewhere, the changes are spatial in nature. In a draft of the chapter 'Farewell to Lórien', Celeborn details that the River 'will pass through a bare and barren country before it flows into the sluggish region of Nindalf, where the Entwash flows in. Beyond that are Emyn Rhain the Border Hills ...' (1989, 281); a later rewriting amends this to 'the River will pass through a bare and barren country, winding among the Border Hills before it falls down into the sluggish region of Nindalf ...' (1989, 281). Subsequent iterations of the map represent this change: originally, the map shows a cluster of hachured mountains beyond the Entwash labelled the Border Hills; a small square of paper inserted onto the map redraws the area, erasing the Border Hills and replacing them with an area labelled the Brown Lands. These changes not only visualize the emerging narrative, they also reveal the map's emphasis on accuracy of distance and direction.

Despite these and many other alterations to the First Map, it remains – as Christopher notes – a decidedly aesthetic as well as practical document. Hachures are used for depicting the various mountain ranges, there is a detailed coastal outline, and colour is symbolically used, including small green treetops

---

[10] See *The Art of The Lord of the Rings* for further examples.

in Mirkwood, blue rivers and red hachures around Mount Doom. In 1943, Christopher redrew this map along with 'A Part of the Shire'.[11] Although this map no longer exists, Christopher describes it as 'a large elaborate map in pencil and coloured chalks' (1989, 299), which stayed largely faithful to the First Map upon which it was based, with the exception of a pictorial style used for mountains and hills. Evidently, Christopher was opting for the pictorial form found in his father's *The Hobbit* maps and his own eventual published maps for *The Lord of the Rings*, even before the limitations of publication were introduced. This suggests that, although Christopher would have been motivated by publication restrictions, his pictorial representation was also partly an aesthetic decision, based on his father's previous maps and older sources, whose visual style he wanted to convey.

A second draft map of Middle-earth is undated, but was made when Tolkien was writing Book V of *The Lord of the Rings*. The map focuses on the southern portion of Middle-earth, and indeed covers much the same area as the map of Rohan, Gondor and Mordor. It is of a much smaller scale, however: it is ruled in squares of 2 centimetres, with each square representing 100 miles (J. R. R. Tolkien 1990, 433). The map makes some use of colour, particularly using blue for the rivers and the coastlines, as well as red for certain annotations; however, the map's main emphasis is on the mountains, which it represents with intricate contour lines and opaque black shading. A final draft map was made after Tolkien finished writing Book VI of *The Lord of the Rings*, in September 1948 (Hammond and Scull 2015, 199). The map was divided across two sheets: a northern portion extending from the Northern Waste to the Falls of Rauros, and a southern portion stretching from Rauros to Far Harad. The maps combine a number of techniques used previously, including a compass rose in the corner and a scale of 2 centimetres to 100 miles; hachures and contour lines to depict gradients – interestingly, the northern portion makes greater use of hachures particularly for the Misty Mountains, while the southern portion almost exclusively uses contour lines; coloured pencils for the forests and rivers; and a more elaborate

---

[11] Although it is generally known that Christopher aided his father before the publication of *The Lord of the Rings* in redrawing the maps, Christopher was in fact a key creative force throughout the process. Christopher drew maps both of the Shire and Middle-earth in 1943, and in various letters, Tolkien laments Christopher's absence during the war: 'I wish I had you here ... completing the maps and typing' (2006, 79); 'my youngest boy ... was carried off last July – in the midst of ... doing a lovely map ... ' (2006, 86); 'He was dragged off in the middle of making maps ... ' (2006, 112); 'chapters went out to Africa and back to my chief critic and collaborator, Christopher, who is doing the maps ... ' (2006, 118). It is important to note, therefore, that Christopher did not merely take over the illustration process at the very end; rather, his maps developed alongside the writing of *The Lord of the Rings* and alongside Tolkien's own map-making efforts.

and at times red script for the larger and more important place names. New and notable elements in the map include the use of contour-like lines to depict the coastline and sea, and notes historicizing the landscape, such as 'Here was of old the Witch-realm of Angmar' and 'South Gondor, now a debatable and desert land'.

In 1953, when Christopher redrew the general map, he therefore undoubtedly drew upon this map as well as the original First Map. Although the map is unarguably a product of Christopher's creative efforts, the debt it owes to Tolkien's own work cannot be denied. The style of the published map very much emulates that of the Wilderland, in the pictorial, individualized depiction of trees and the face-on rendition of the mountains. The mountains of Mordor also strongly resemble an aerial sketch Tolkien carried out of Ered Lithui and Ephel Dúath. This map features a compass rose in one corner and a scale in the opposite corner. A later redrawing of the map made for *Unfinished Tales* in order to incorporate new locations and to correct defects in the original map also features the compass rose and scale; above the scale is the title 'The West of Middle-earth at the End of the Third Age'. This map also features contour-like lines in the sea, reminiscent of those first seen in Tolkien's 1948 map.

Wood cites the draft maps of Middle-earth in his discussion of the process of map-making, arguing that typically, maps as individual objects do not 'grow' or develop, but are rather informed by systems and practices that change over time. The exception to this, Wood explains, is literary map-making, which develops alongside the world and fictional cartographic practices that it depicts: examining the First Map and the ways in which places were renamed, distances recalculated, and entire territories erased, pasted over and redrawn, Wood admits that

> [h]ere we see not just growth and decay, but also development, for what J.R.R. Tolkien did was to continuously differentiate, articulate and hierarchically subordinate the parts of the Middle Earth [sic] he was creating ... *interactively* ... with this map; so that history appears here, in the way the map takes as given certain aspects of Middle Earth [sic] previously worked out, even as it – precisely – generates others ...
>
> (1992, 30–31)

Wood's discussion of the map's hierarchical control over the broader text is only one aspect of how the Middle-earth map creates and encodes narratives of power. Unlike Thror's Map, which combines medieval understandings of navigational cartography with other forms of medieval cartographic structuring, such as the literal marginalization of dangerous areas, the Middle-earth map's structure instead illustrates modern cartography's emphasis on accuracy and representation, which embodied a desire to master and control the landscape. The topography of the map is framed by a compass rose and a scale bar marked at 50-mile intervals, indicating its concern with accurately conveying distance and direction. Moreover, the Middle-earth map is heavily focused on toponymical representation: from large-scale territories, forests and mountain ranges to small-scale villages, towers and paths, the map is saturated with place names. Discussing 'A Part of the Shire', which has a comparable density of place names, Ekman argues that 'the map subjugates the landscape, brings it under control ... ' (2013, 50); the Middle-earth map similarly displays an intricate knowledge of the land that places the map in a position of epistemological control over the landscape.

This practice very distinctly draws on the traditions of post-Enlightenment cartography in Britain, where surveys and scientific measuring using new technologies were employed in order to create maps that could efficiently disseminate large amounts of information and thereby claim a complete knowledge of the landscape depicted. Both in the Ordnance Survey and military maps, the innate incompleteness of cartography was disguised by the presentation of objective details such as scaled distance and marked place names; nevertheless, these maps remained, as Wood and Fels argue, socially and politically constructed texts that manifested their ideology of knowledge through their very supposed objectivity. By imitating these techniques, noticeable both in Tolkien's meticulous attempts to maintain consistency of distance between the narrative and the map, and in the map's level of topographical and toponymical detail, the Middle-earth map similarly aims at a supposedly objective and complete reproduction of the world that allows Tolkien to regulate his sub-creation while simultaneously enabling considerations of the diegetic tension between the cartographic image and the independent reality of the natural world.[12]

This modern conceptualization contrasts heavily with the map's overall medievalist aesthetic identified by Padrón, Fimi, Hammond and Scull. Its iconography rejects the abstraction that would be expected from a map that is so intently focused on fidelity of distance and scale, drawing instead from the illustrative techniques of medieval cartography. The forests are represented by tight clusters of individually demarcated trees, and in certain areas such

---

[12] This is discussed in much greater detail in Chapter 2.

as the Trollshaw or Nan Elmoth, the species of trees are visibly different: some are shorter, rounder and deciduous-like, while others are taller and pointed like conifers. While modern maps such as the Ordnance Survey certainly use pictorial markers for trees, they employ a repeated standardized symbol to indicate wooded areas; the Middle-earth map tends instead to the individualized depiction of medieval maps. In certain areas, man-made structures are shown in a similar, illustrative way: much as the mappae mundi and the Gough Map showed houses, churches and castles face-on, the towers of Barad Dûr in Mordor, Dol Guldur in Mirkwood and Gondolin in Dorthonion are shown in profile, rather than from above. The many mountain ranges of Middle-earth also take inspiration from this style of medieval maps; although Christopher's depiction is more sophisticated than the slightly misshapen examples on the Gough Map, he nevertheless also depicts mountains as a series of peaks, moving away from the contemporary, contoured representation of relief seen in the First Map.

At the same time, unlike Thror's Map, the Middle-earth map does not make use of non-topographical illustrations such as dragons, spiders or manicules to illustrate political or historic concerns. Although its use of illustrative symbols is aesthetically medieval, it nevertheless tends towards modern cartography's emphasis on the stable and topographical, dismissing the distinctly subjective narrativization that these additional symbols would bring. The map is thus a definitive embodiment of Eco's theory of the pseudomedieval: rather than an authentic reproduction of medieval practices and ideologies, it instead depicts a 'fantastic neomedievalism' onto which contemporary ideas can be projected (Eco 1987, 63). In this case, these contemporary ideas revolve around the ideologies of modern cartography and the ways in which it creates and maintains a hierarchical relationship between the map-maker and its subject based on knowledge and accurate representation. The Middle-earth map's large scope allows it to speak to these broader cultural questions; the power dynamics engrained within the map become applicable to Tolkien's wider sub-creation, as opposed to Thror's Map, which was defined by its specificity as an itinerary map. The commonly accepted medievalism of the Middle-earth map thus requires nuancing: the contemporary issues that Tolkien interrogates throughout his world-building – from the environmental concerns addressed in Chapter 2 to the critique of power politics and modern imperialism in Chapter 4 – necessitate the modern conceptualization of cartography that predominantly informs the map, with the medieval in this case acting largely as an aesthetic overlay.

Map V: Map of Rohan, Gondor and Mordor

In 1948, while writing Book VI of *The Lord of the Rings*, Tolkien also made a draft of his map of Rohan, Gondor and Mordor. This map covers the terrain where the primary events of this book take place, stretching from the East Fold in Rohan across to the region of Nûrn in Mordor. The map was drawn on 2.5-mm-ruled graph paper, with red squares ruled over the top every 100 mm. A note at the top reads: 'Small Scale 100 miles = 20 mm. (1 mm. = 5 miles) | Large Scale ×5: 100 miles = 100 mm. 1 mm = 1 mile'. Hammond and Scull hypothesize that the small-scale map Tolkien refers to here is his general map of Middle-earth drawn at the same time, which focuses on the southern portion of the land, with the map of Rohan, Gondor and Mordor showing the area in five times more detail (2015, 205). The map was used to track Frodo's and Sam's journey from the Falls of Rauros down to Mount Doom, with each day of the journey marked as a number alongside the trail. The graph paper and the large scale of the map enabled Tolkien to trace the journey as accurately as possible through the landscape of Middle-earth. The map's emphasis on accuracy is maintained in its mode of representation. There are no stylized trees or mountains here: the terrain is entirely depicted using intricate contour lines, including the sea at the Bay of Belfalas. Colour is used to pick out features, including blue for the network of rivers, red for the region's names and purple for the beacons of Minas Tirith, which are also numbered. The map is also toponymically very detailed, with areas, settlements and rivers all labelled.

Despite the level of detail, this was clearly a working map, indicated by the note in the top left-hand corner that 'Entwash is too far east', and required changes to bring it up to publishing standard. In a 1954 letter to Allen & Unwin, Tolkien remarked that '[a] map of the Gondor area is perhaps the most urgent. I am hoping to get my son Christopher to produce one from my drafts ...' (Tolkien 2006, 185). A few months later, however, Tolkien wrote in a letter to Katherine Farrer that Christopher was 'too overwhelmed to help with maps', and attempted to redraw it himself (2006, 208). This proved difficult, as detailed in a letter to Rayner Unwin from 1955: 'The map is hell! I have not been as careful as I should in keeping track of distances. I think a small scale map simply reveals all the chinks in the armour – besides being obliged to differ somewhat from the printed small scale version, which was semi-pictorial ...' (2006, 210). Tolkien and Christopher eventually finished the large-scale map together, with Tolkien 're-scaling and adjusting' the measurements and Christopher redrawing the entirety of the map over twenty-four hours (2006, 247). The redrawn map is unique, as it is the only one of Christopher's maps to retain his father's contour lines, albeit simplified for ease of printing. This was potentially in order to visually distinguish the map from the Middle-earth map: in a letter to Rayner Unwin, Tolkien emphasized that the larger-scale map needed to differ from the small-scale one, possibly to avoid repetition and to offer the reader a new perspective on Middle-earth. Nevertheless, although the map does eschew the typical pictorial depiction of relief, it does not avoid pictorial representation altogether. Christopher added the by now iconic clusters of trees for Firien Wood and Drúadan Forest, symbols of grass for the Wet Marshes and a small tower for Barad Dûr.

Like the Middle-earth map, the map of Rohan, Gondor and Mordor is conceptually a modern map, yet in this case, its modern structure is visually represented by contemporary representational techniques. The inclusion of a compass rose and scale bar once more signal the importance of accuracy, which is reinforced by Tolkien's use of contour lines to depict relief. Interestingly, however, the contour lines do not necessarily convey more detailed or precise information about the terrain of these territories: the lines correspond exactly to the illustrated mountains in the Middle-earth map, including details such as the lone peak of Emyn Arnen on the border of Mordor, and the valley of Udûn, depicted on the map of Middle-earth as a gap between shaded mountain peaks, and on the map of Rohan, Gondor and Mordor as a blank space between contour lines. The purpose of the contour lines is therefore not to depict the lands of Middle-earth in noticeably greater detail, but rather to give the appearance of doing so, and to present the map as more accurate. This was partly a paratextual choice – Tolkien's letter to Rayner Unwin makes it clear that he was attempting to aesthetically differentiate the two maps – however, the choice also inevitably embeds modern cartography's preoccupation with control over the landscape to a greater visual degree than the general map of Middle-earth.

In particular, the map of Rohan, Gondor and Mordor closely aligns with the military maps of the early twentieth century, a connection that is reinforced by the map's clear hermeneutic attention on war: not only does it centre on the areas of Middle-earth where fighting takes place in *The Return of the King*, but the sublabel 'Battle Plain' beneath Dagorlad, as well as the careful labelling of the enemy territory of Mordor, suggests this map could be used for strategic purposes. The categorization of areas from a military perspective is reminiscent of Tolkien's own the First World War trench map, where annotations indicated the reader's engagement with the politics of the landscape. This connection between the map of Rohan, Gondor and Mordor and post-Enlightenment military mapping serves several functions. First, it demonstrates the influence of modern cartography on parts of Tolkien's corpus, aesthetically, structurally and functionally. Second, it reinforces the concept of the map as a sociopolitically constructed document that assimilates and reflects back the ideological conditions of its production. Third, it emphasizes the connection between military power and violence and the land, and the ways in which modern cartography textually embodies the physical control over land that military activity secures. As Lieutenant Colonel E. M. Jack claimed, the map becomes part of the arsenal of war (Chasseaud 2013, 10), and the map's accuracy is exploited as a tool for gaining control over land and the people who live in it. As political conflict forms a central aspect of the legendarium's narrative,[13] the ability of Tolkien's maps to embody modern cartography's aspiration for control – as in the military aesthetic of the map of Rohan, Gondor and Mordor – is paramount.

These five groups of maps only form half of Tolkien's overall cartographic corpus; however, even this limited sample clearly demonstrates Tolkien's engagement with the historical role of maps in creating, embedding and enabling ideological and political ideas. Tolkien draws on the medieval and modern periods to varying extents in different maps, thereby forming a cartographic practice that is dependent on both periods stylistically, and, more crucially, conceptually. By replicating the techniques that the maps from these periods employed, Tolkien situates those of his sub-creation within a historic tradition of political cartography that emphasizes the inherent nature of maps as 'value-laden images' (Harley 1988, 278). This not only enriches his maps from a paratextual perspective, signalling political and cultural contexts to his external readers, it also opens them up to reflecting the broader political concerns that Tolkien examines throughout his legendarium, further emphasizing how cartography is inextricably enmeshed with its sociopolitical context, even if this context is fictional. The following three chapters will build on this positioning of maps within historical ideologies by considering how Tolkien's maps work alongside the text to articulate narratives of the environment, deep time and geology and power politics and imperialism, in order to demonstrate the multiple methods Tolkien employed in order to engage his work with its contemporary sociopolitical context.

[13] This is discussed in depth in Chapter 4.

2

Environment

Before we can return to maps, we must first turn to land, for it is from our conceptualization of the landscape – its non-human tangibility, subjectivity and power – that our attempts to fix it in place emerge. Tolkien was, of course, a famously committed nature-lover; in 1972 he wrote to the editor of *The Daily Telegraph* that 'in all my works I take the part of trees against all their enemies' (2006, 339), and later lamented in another letter that a potential film-maker for *The Lord of the Rings* was not 'interested in trees: unfortunate, since the story is so largely concerned with them …' (2006, 210). His legendarium has often been read as an analogue to what he perceived as the depraved effects of industrialization on the peaceful green landscape of his homeland and beyond; the scorching fires of Mount Doom finding modern resonance in the chimney stacks and smoke that rose up over the Midlands of his childhood.

For ecocritic Timothy Morton, Tolkien's approach to nature – as has been so popularized by these broadly conceived analogies and constantly reproduced photographs of the author peaceably sitting by trees – is emblematic of a vague nostalgic interest in a kind of past natural harmony, and a problematic yearning for a return to 'nature' as an environmentalist utopia. Nature, Morton argues, has become so romanticized and abstracted as a concept that it has itself become an obstacle to proper ecological practice, as environmental writers are entirely distanced from the political, material and ontological realities of the environment for which they supposedly advocate. Morton cites Tolkien as an example of this harmful ecological approach; framing the Shire in *The Lord of the Rings* as a 'world-bubble' and organicist fantasy, Morton argues that Tolkien promotes a myopic and idealized engagement with nature that refuses to acknowledge the wider world of global politics. 'If ever there was evidence of the persistence of Romanticism', claims Morton, 'this is it' (2007, 97).

While certain descriptions of the Shire build on a pastoral and indeed – as Morton defines it – arguably Romantic view of nature (although I counter that

Tolkien's general narrative thrust is a maintained subversion of the deliberate idyll with which he introduces his novels), this chapter will contend with the broader, implicit categorization of Tolkien's work as persistently Romantic, idealized and depoliticized. Tolkien's legendarium is, as this chapter will show, fundamentally grounded in not just an engagement with but a studied and forceful critique of the burgeoning ecological crisis that was already starting to suggest itself in the early and mid-twentieth century. Moreover, when considered alongside ecocritical, and specifically posthuman, lines of thought, the legendarium reveals itself as a sharply political text that well understands the frightening consequences of human power over the non-human. Tolkien's mapmaking and his making more broadly – his sub-creation of a new, imagined nature just as vulnerable to continued exploitation, degradation and ruin as our own – act as a prism through which our own world's harmful exercise of power over the non-human can be focalized, understood and resisted.

It is perhaps worth noting here that the intention when placing Tolkien in conversation with posthuman critical texts, which contribute to a theoretical field that well postdates his death, is not to retroactively apply this school of thought to his writing, or to anachronistically pigeonhole him as a particular kind of environmental thinker. Rather, I hope to draw attention to the ways in which his legendarium, long before these ideas were theorized and concretized within our philosophical paradigms, engages in a similar critical environmentalism that extends beyond an adulation of 'nature' into a fearful consideration of our long-established exploitative relationship with the non-human world and its wide-ranging implications. While the specificities of ecocriticism shift across the spectrum of the environmental humanities, they are all of them preoccupied by the same anxieties of cause and effect (our collective, thoughtless actions, their inescapable outcomes) which sit heavy across Tolkien's literature. Writ large across both corpora is the inevitable doom to which the hierarchies and binaries we have constructed around ourselves and the non-human will eventually lead if we do not radically alter our conceptualizations of both.

Beginning with a very brief examination of the touchstones of environmental posthumanism that can help to illuminate our understanding of Tolkien's own approach, this chapter will emphasize how Tolkien's sub-created cultures, much like our own, frequently position the non-human world as something to be overcome and defeated. It considers how Tolkien draws attention to the frequently exploitative and controlling act of cartography and to portents of irretrievable environmental harm in order to advocate for an alternative where the non-human is valued not for its aesthetic or practical possibilities, but as an independent subject that holds its own vitalities and vibrancy. By imagining a world where the environment is empowered beyond what our rational realism can allow for, Tolkien makes from his legendarium a form of ecological protest, drawing on the possibilities of the fantasy genre to re-enchant nature and resist the control that mapping represents. I want to consider Tolkien's legendarium not merely as an elegiac lament for the idealized nostalgia of the 'natural world', but as the response of an author writing after the mass industrialization of the nineteenth and twentieth centuries and the environmental toll of the world wars, as an author writing – although he was not aware of it himself – on the brink of the Anthropocene.

Navigating the Human, Non-human and Posthuman

Coined by atmospheric chemist Paul J. Crutzen in 2000, Anthropocene is a speculative name for our current geological epoch, in which the human – or Anthropos – is the key affecting factor in our planet's ecological and geological systems. Intended to follow the relatively peaceful Holocene epoch that began after the most recent Ice Age, the Anthropocene has yet to be ratified by official geological societies and its Golden Spike, or starting point, is still a mutating, shiftable thing: some date it to the invention of the steam engine in 1776 and the subsequent Industrial Revolution (rooted in and enabled by forced slave labour in Britain and abroad), others to the Great Acceleration that followed the Manhattan Project and the detonation of nuclear bombs unleashed on the planet and its people. What is clear, however, is that the unprecedented ecological damage inflicted by humans is now written into our geological strata: non-degradable technology embedded in rocks, the oceans acidifying beyond help.

The Anthropocene is a crystallization of the terrifying, destructive ways in which a certain collapse of boundaries has been allowed to take place. This is a collapse predicated not on the productive possibilities inherent in communion, care and an expanded understanding of the human and non-human self, but on the continued, solipsistic intervention of the human onto the non-human: its environments, its temporalities, its value. There are numerous factors that have led to the Anthropocene – global industrialization, our apparently indispensable reliance on fossil fuels, the advent of capitalism located within the horrors of the slave trade – yet that the Anthropocene has been allowed to take place is due, very broadly speaking, to a centuries-unchanged relationship between the human and non-human: one that sees the relationship between the two as a hierarchical, extractive binary.

The work of ecocriticism, and environmental thought more broadly, has been one of a continued and increasingly radical decentring of the human within our conceptualization of the human and non-human spheres. Ecocriticism recognizes the sticky mess of the Anthropocene, the frankly unnatural state of geological and ecological affairs we have landed ourselves – and the planet – in, and tries to imagine new possibilities. What if we understand that our needs and desires do not define the non-human? What if we understand that it has desires and intentions beyond us? What if we could understand it as both a separate, independent entity – with all the concomitant subjectivity that would entail – and as something with which we are profoundly entangled? What if we could sit within a better, emancipatory collapse of boundaries – and wait?

This practice of decentring the human within the field of environmentalism traces to the deep ecological movement, an environmental philosophy founded by Norwegian philosopher Arne Naess in the early 1970s that was greatly influenced by the godmother of the modern environmental movement Rachel Carson and her seminal text *Silent Spring* (1962). Rather than the anthropocentric attitude Naess observed in 'shallow' ecology, an approach based in anxieties around resource depletion and the human cost of pollution, deep ecology advocates for a biocentric approach, one that understands the intrinsic value of nature beyond any aesthetic, material or nurturing use it may have for humans (Sessions 1995, xii). For deep ecological practice to work, the long-established binary between the human and non-human and the ontological and epistemological rupture that has kept the two categories opposed and the non-human inferior and vulnerable to the human must be interrogated. Eerily familiar are the similarities between deep ecology's condemnation of the harmful, extractive consequences of viewing the non-human through a use–value lens and critical cartography's critique of power and knowledge. Cartography, after all, is the ultimate instrumentalization of the non-human, rendering the landscape in its entirety into a tool for human consumption.

For ecofeminist critic Val Plumwood, the exploitation inherent in a human–non-human dualism can only be understood through broader dichotomous structures of oppression, such as systems of patriarchal power that oppose the feminine and masculine and subjugate the former within the latter. These related dualisms, Plumwood argues, are entrenched within a Western, post-Platonic emphasis on rationalism as the defining characteristic of the human, with nature by contrast portrayed as unthinking and compliant (1993, 5). The much-needed deconstruction of the binary therefore rests on eschewing reason entirely for a broader definition of mind-like qualities. The non-human, Plumwood argues, is, if not rational, then certainly intentional. In acknowledging this intentionality – the consciousness and deliberation and teleological life goal of non-human behaviour that evinces its mind-like capacities – there is ground for continuity between mind and nature, a continuity that is suddenly, wonderfully alive with heterogeneous possibilities (1993, 134). For Plumwood, deep ecology does not go far enough in either acknowledging the broader structures of hierarchical power that inform the human/non-human hierarchy, or in properly conceptualizing the new entanglement between the two hitherto segregated parties. Her critiques are echoed in Morton's scepticism of our cultural obsession with a return to nature. For deep ecology – a really deep ecology – to work, Morton argues, the very idea of nature and the distance created between us and it must be let go (2007, 204). The shift from anthropocentrism to biocentrism is not enough: a greater collapse of boundaries is needed.

Both Plumwood and Morton are bumping up against the edges of a posthuman line of environmental inquiry; an inquiry that, rather than trying to build a bridge between the domains of the human and non-human, comes to question the boundaries of these domains entirely. It is worth clarifying here precisely what we mean by the posthuman, a markedly slippery term which reaches into the fields of cybernetics, science studies and radical feminism alike. We do not mean transhumanism, which examines how humans might harness technology and nature in order to expand and enhance the natural boundaries of human capability: an intellectual investigation that would certainly prolong rather than curtail the Anthropocene. Cybernetics, and more precisely the figure of the cyborg as famously theorized by seminal posthumanist Donna Haraway, is crucial to posthuman environmentalism, but only in as far as it provides a cipher for understanding the mutability of the supposedly fixed boundaries between human and non-human and the infinite potential for entanglement. For a neat dictionary definition of the posthuman, we might turn to Bruce Clarke and Manuela Rossini, who describe it as 'states that lie before, beyond, or after the human, or into which the human blurs when viewed in its essential hybridity' (2017, xiv). What is key to the posthuman is a troubling of the categorization – and its concomitant culturalization and superiority – of the human, and thus the non-human with it. As Rosi Braidotti argues, within the deconstruction of an anthropocentric perspective lie entirely new possibilities of contemporary subjectivity and subject formation that go far beyond humanist imaginations (2013, 58).

For Haraway – her of cyborg fame – there exist energizing environmental implications of such posthuman subjectivities. Her approach is notable in its productive rather than destructive qualities: the hierarchical extractive binaries that have long defined the boundaries between human and non-human must be done away with, but this is framed through a process of kin-making, specifically 'oddkin'-making rather than 'godkin'-making; that is to say, forming networks of interdependence that unsettle pre-existing, ordered concepts of relation (2016, 2). Connection, responsibility and recognition are formed across and beyond genealogical and biogenetic divides, resulting in 'compost piles' (2016, 4): inherently reproductive and creative slurries of coexistence. Central to Haraway's argument is her eschewing of the terms Anthropocene and Capitalocene (a variation on the Anthropocene that points to capitalism as the epicentre of our environmental crisis) for Chthulucene, a speculative era that centres the chthonic: the inhuman, non-human, more-than-human. As Haraway explains, human beings are no longer the central actors in the Chthulucene; agency is given (back) to other beings and hierarchical orders are unravelled and rewoven so that human beings are merely a part of broader biotic and abiotic powers. 'We are', Haraway explains, 'at stake to each other' (2016, 55).

Contained within the Chthulucene is a wholesale disruption of our understanding of the non-human and our relation with it. Gone are solipsistic forms of human/non-human dependence, be they extractive or condescendingly conservationist; gone too is the ardent anthropocentric belief that agency and subjectivity are but the realms of the human. Haraway's brave new world suggests not only an epistemological reconfiguration – similar in ways to Plumwood's expansive reimagination of mind-like activity – but an ontological one, one that (as we will examine below) has decidedly material implications on how we perceive the non-human.

Tom Bombadil and the Non-human

Admittedly, this may all seem very theoretical, very abstracted, very *modern*. But it is, I believe, vital, to recognize that Tolkien was dancing around many of these same ideas in his Middle-earth writings: a fundamental mistrust of the separation between the human and non-human, a disgust of this separation's proclivity towards hierarchy, a fascination with decentring the Anthropos and recognizing nature as a kinetic force that begins to collapse the ontological and phenomenological boundaries between the human and non-human.

Tolkien's writing touches on some of the foundations of deep ecology in his dislike of instrumentalism, his fervent belief in the value of nature beyond human need. But, as we shall see, he also goes beyond this towards philosophical queries into the boundaries between the human and non-human, stressing a non-human intentionality and a wild, chthonic vibrancy that disrupts the boundary that cartography, among other things, attempts to fix around the human and non-human. It is through sitting with these ecocritical arguments that we can appreciate the specific ways in which Tolkien was engaging with ecological concerns; a radical approach that goes far beyond the love of trees with which he is so associated.

Before we begin to examine the role of maps in attempting to fix this hierarchy, and the ways in which the non-human in Middle-earth resists, it is worth briefly looking at a case study from *The Lord of the Rings* that epitomizes Tolkien's environmentalist ethics and the attention he pays to the non-human's intrinsic value. The elusive figure of Tom Bombadil stands somewhat strangely apart in the legendarium, existing as he does on the fringes of the narrative both in terms of interior mythology and sheer presence within the story, to the extent that Tolkien himself refers to his character as an 'intentional' enigma (2006, 174). Yet his separateness from the broader political, social and historic structures of Middle-earth allows him to embody attitudes towards human/non-human relations that lie beyond the inherently hierarchical position that almost all other beings in Middle-earth take, presenting an alternative and radical response to tendencies towards anthropocentrism and instrumentalization that lays a foundation for understanding new forms of non-human conceptualization in Tolkien's writing.

This reading is one that is largely supported by Tolkien's own interpretation of his character, which he elucidates in two letters: one sent in 1954 to Naomi Mitchison and the other in the same year to Peter Hastings. In the former, Tolkien makes explicit the danger of human power over the non-human – so far, so good – yet also makes explicit the antithesis of it: the radicality inherent in decentring the anthropic self from the non-human, and the potential beauty in comprehending things 'for themselves':

> The story is cast in terms of a good side and a bad side, beauty against ruthless ugliness, tyranny against kingship, moderated freedom with consent against compulsion that has long lost any object save mere power, and so on; but both sides in some degree, conservative or destructive, want a measure of control. But if you have, as it were taken 'a vow of poverty', renounced control, and take

your delight in things for themselves without reference to yourself, watching, observing, and to some extent knowing, then the question of the rights and wrongs of power and control might become utterly meaningless to you, and the means of power quite valueless. It is a naturally pacifist view, which always arises in the mind when there is a war.

(2006, 178–79)

This letter is crucial both in terms of understanding the uniquely subversive position that Tom Bombadil holds in the legendarium, however briefly he might appear, and also in demonstrating the necessity of complicating the moral binary that the legendarium otherwise revolves around. By underlining the control that both sides strive for, whether conservative or destructive, Tolkien irrefutably demonstrates inherent dynamics of power that define historic relations between the human and non-human, whether intentionally damaging or not, whether from the 'bad guys' or the 'good'. This argument is further elucidated in the second letter, also on the topic of Tom Bombadil, which explains:

He is master in a peculiar way: he has no fear, and no desire of possession or domination at all. He merely knows and understands about such things as concern him in his natural little realm ... he is then an 'allegory', or an exemplar, a particular embodying of pure (real) natural science: the spirit that desires knowledge of other things, their history and nature, because they are 'other' and wholly independent of the enquiring mind, a spirit coeval with the rational mind, and entirely unconcerned with 'doing' anything with the knowledge: Zoology and Botany not Cattle-breeding or Agriculture. Even the Elves hardly show this: they are primarily artists.

(2006, 192)[1]

---

[1] It is worth noting here the slight contradictions of terminology between Tolkien's writing and the environmental, critical cartographical and Foucauldian theories that we have examined until this point. In this letter, Tolkien praises having knowledge over nature, a problematic notion when considered in Foucauldian terms; however, the rest of the letter makes clear that this knowledge is not contingent on control or definition, but rather an acknowledgement of the alterity of nature – a term that again might conflict with an understanding of ecocritical and postcolonial terminology but in this case simply means conceding nature's difference as Plumwood lays out. Tolkien's definition of rationality is similarly built on understanding the complexity of the natural world and humanity's fractional place in it, rather than a denial of that quality to the non-human, as per Western philosophical thought. In *The Lord of the Rings*, meanwhile, Goldberry refers to Tom Bombadil as 'Master' of nature (2008a, 163), a concerning framework that is belied by the rest of her speech which highlights the independence of their non-human surroundings. In this letter, Tolkien makes explicit his definition of this mastery: it does not comprise either possession or domination, which are typically the criteria for mastery, but rather of knowing and understanding the surrounding world. This chapter will use these terms not per Tolkien's own unique definitions but per that of the theory we have outlined, in order to make this study as broadly applicable as possible.

Tolkien's argument here may as well have been lifted from the pages of a deep ecology treatise. He maintains a binary between the human and non-human, but one that is predicated on understanding the non-human as apart from oneself, and in resisting the solipsism that would lead to acts of domination, subjugation and extraction. His scepticism of 'doing' anything with knowledge of the non-human demonstrates the slippery slope that such an approach can take: Bombadil resists Cattle-breeding and Agriculture in much the same way, and for much the same reasons, that he resists domination and possessiveness. The pursuit of instrumentalizing such knowledge is, Tolkien warns, deeply vulnerable to dangerous power.

In *The Lord of the Rings*, Tom Bombadil's chapters advocate for the necessity of such an ecocentric, arguably deep ecological approach, while also beginning to interrogate subjectivity as the realm of the human and thus blurring the boundaries of human and non-human definitions. Tom Bombadil first appears in the tale in order to rescue Frodo, Sam, Merry and Pippin from the grips of Old Man Willow and, shortly after, proceeds to tell them stories not just of the flora and fauna of the forest, but of its very selfhood. '[E]vil things and good things, things friendly and things unfriendly, cruel things and kind things, and secrets hidden under brambles' make up his tales (2008a, 170), depicting the richness of a world imbued with intentionality and subjectivity. Listening to Tom Bombadil, the sheltered hobbits begin to understand the world 'apart from themselves'. Their conceptualization of themselves as masters, or indeed as protagonists, of the Old Forest begins to shift, reminding them that they are strangers in a place 'where all other things were at home' (2008a, 170). It is particularly striking that it is hobbits – famously creatures defined by home, comfort and belonging in familiar surroundings, and associated with the romanticized pastoral of the Shire in much scholarship – who are destabilized within the order of things. Their position as interlopers in an ecosphere that exists richly apart from them is all the more disorienting, and it disorients with it the reader's own sense of surety over the non-human.

The richness of the non-human is stressed by Tom Bombadil's contradictory descriptors: good and bad, friendly and unfriendly, kind and cruel. Old Man Willow may be described as malicious, but this ambivalent portrayal of the non-human works to resist an idealized, Romantic image of nature; a tendency that Morton explains often frustrates truly emancipatory ecological thought. In telling the hobbits about Old Man Willow and the Old Forest, Tom Bombadil lays bare 'the hearts of trees and their thoughts' (2008a, 170), a turn of phrase made distinctly unmetaphorical by its specific unpacking of the subject position of these non-human creatures: their hatred of the things that 'gna[w]', 'hac[k]' and 'bur[n]' (2008a, 170). Tolkien's language interweaves the plant-like and mind-like: the Old Forest is the last remnant of 'forgotten woods' who 'remembers' better times, and the trees within it contain 'rooted wisdom' (2008a, 170). The Old Forest has subjectivity, memory and intentionality, not despite but precisely *because of* its non-human nature. In recognizing the depths of this chthonic alterity, a non-extractive relationship with the non-human is both brought to the fore and its absence in other human/non-human relationships in Middle-earth thrown into relief. Despite his brief, anomalous narrative intrusion, Tom Bombadil is vital within the ecology of Middle-earth exactly due to his anomalous nature. His character shows us what might be possible if we reimagine the human and non-human – if we begin to tell new stories.

Mapping the Human and Non-human in Middle-earth

As introduced above, the sought-after 'measure of control' that Tolkien details in his letter to Mitchison lies at the heart of the binary that exists in his sub-creation, including in his maps. His Middle-earth texts centre on humans and humanoid creatures living, often uncomfortably, in relation to their environment, depicting a version of the dualism and hierarchy that structured his own world and its concomitant degradation of the environment. Tolkien's comment that both sides to 'some degree' want this control indicates a spectrum of power over the environment, from those who simply practise harmful systems of knowing in order to frame the natural world as apart and inferior, to those who enact complete mastery over their environment through active damage.

Cartography is one of the most totalizing instrumentalizations of the non-human, rooted as it is in the Western post-Renaissance project of rationalizing the non-human world by offering a purportedly correct, logical and objective representation that offers knowledge and control. The very act of mapping Middle-earth speaks to an attempt to understand, categorize and dominate the non-human; much as in the primary world, maps are used as a way of familiarizing, negotiating and mastering unknown landscapes. The map makes the landscape readable through the medium of cartography; the nature of the map as a textual object, meanwhile, which is passed around and read by various people, means that it is not only the original cartographer who possesses this understanding; rather, every user of the map can come to 'read' the landscape, using the map as a tool. This transfer of knowledge positions the natural world as something that can be rationalized and condensed onto a page, subjugating it within a power hierarchy that negates its own, intrinsic value.

We see this when Thorin is first offered Thror's Map by Gandalf, who explains that the map comes with a key; his framing of the two items as a single unit draws parallels between their functions and foreshadows the map's ability to 'unlock' the mysteries of the non-human world depicted on the map. At first, however, Thorin dismisses his need for the map, claiming to know where Mirkwood is, and the mountain 'well enough' (2008b, 26). Thorin's scepticism is, perhaps, understandable: at first glance, Thror's Map does not seem to give any particular topographical detail or indicate any paths that will aid the party in reaching their destination. Yet, as Gandalf points out, the map offers a new level of insight: the runes on the left-hand side point to the location of a secret entrance that will allow the dwarves access. Indeed, this is largely how Thror's Map is used. Rather than the presentation of accurate, scaled landscapes and roads, the value of the map lies instead in its instructive writing: the moon letters that Elrond discovers offer a way into the mountain which no mere topographical replica could. The reliance on the map to navigate and penetrate the mountain, even one that was home to the dwarves, renders the landscape a difficult and disconnected entity, one that must be mapped, read and decoded in order to be overcome.

The ability of the map to strengthen the diegetic reader's understanding of their initially impenetrable surroundings is epitomized through Bilbo, whose journey through the wilds of Middle-earth is very much tied up with his formative hero's journey. There have been numerous critiques of the novel within Anthropocene studies as a genre inherently unsuited to subverting the hierarchies between the human and non-human: the traditional novel, after all, is a study of anthropocentrism, of the journey of human characters within the world. Bilbo's position in *The Hobbit* ties into this in quite a literal sense (although, of course, Tolkien himself is critiquing this), in that his process of identity formation is explicitly tied to his increased mastery of the land. His growing sense of self is reflected in his growing confidence reading Thror's Map, and the competence it gives him to read and negotiate his surroundings. Of all the characters, it is Bilbo who engages most with Thror's Map, and who has the most success with it, so that it is he who eventually discovers the secret door, after 'often borrow[ing]' and 'gaz[ing]' at the map, and 'pondering' over its hidden messages (2008b, 261). It is striking that it is the hobbit rather than any of the dwarves who penetrates the dwarvish secrets of the Lonely Mountain: it emphasizes Bilbo's development throughout the text, and highlights how his reading and rereading of the map have worked to give him skills thatch none of the other characters possess. Yet it also suggests that without an external aid, the non-human land of the Lonely Mountain is devoid of meaning and unnegotiable: in neglecting the map, the dwarves are unable to rationalize their environment and remain lost.

The connection between the knowledge that maps communicate and knowledge of and power over the non-human is depicted in the parallels between the characters reading maps and 'reading' their natural environment. By deconstructing and reconstructing the non-human into a socially legible framework, rendering it understandable and subdued, the map enables its readers to enact this same power over their tangible surroundings. Staying with Bilbo, we see that he is unable to accurately analyse his surroundings early on in his journey: he asks, after barely any time has passed from the start of the adventure, whether a mountain range is '*The* Mountain' (J. R. R. Tolkien 2008b, 59), demonstrating both his naivety regarding the enormity of the quest and his inability to grasp the layout and scale of the country. Bilbo also constantly yearns for his home: when climbing the Misty Mountains, Bilbo turns in the direction of his home and reminisces about where things are 'safe and comfortable' (2008b, 70), and Tolkien frequently adds the aside, 'It was not the last time that he wished that!', when Bilbo thinks about being back at Bag End (2008b, 42, etc.), highlighting the foreignness of the environment he is currently in. Bilbo's discomfort – both physically and emotionally – in this unknown, wild landscape creates a rupture between himself and his environment, emphasizing both his sense of displacement and his inability to regulate his surroundings as he is accustomed to in the Shire.

Bilbo's eventual mastering of his surroundings is, therefore, all the more marked and correlates with his increasing occupation with the map. Bilbo's confidence traversing the landscape grows throughout the text, and it is in the Mirkwood chapter that he entirely grasps the workings of his surroundings. That he achieves a level of sensory and cognitive control over the environment is all the more notable given its sylvan nature. Mirkwood Forest is emblematic of what ecocritic Robert Pogue Harrison has characterized the 'hostile opposition between forest and civilization' (1992, 49), arguing that the border between the forest and human habitation is consistently held up as a symbolic boundary between various human–non-human dichotomies, whether that be civilized–uncivilized, domesticated–wild, industrialized–pastoral or known–unknown. Mirkwood is such a forest, standing in contrast to the various homely, ordered habitations and domesticated landscapes in Middle-earth, and instead offering an environment which is '*foris*', or outside the boundaries of civilization. Harrison points to the destabilization of time and space that occurs in forests, arguing that their ability to subvert our most fixed social structures is indicative of their place outside human rationality (1992, 38). Mirkwood very much undermines these categories as Harrison suggests: as they delve deeper into the forest, the travellers' surroundings become increasingly gloomy until the light at the gate becomes but a 'little bright hole' (2008b, 178), suggesting that they have left their ability to calculate and mark time, and thus the very notion of temporal structures and boundaries behind them. This emphasis on the lack of light in Mirkwood continues: the forest becomes a 'dimness' (2008b, 178), and 'everlastingly still and dark and stuffy' (2008b, 180), and the company entirely loses sense of time and place.

Given the subversion of these rational temporal and spatial structures, it is particularly striking that Bilbo manages to successfully navigate and vanquish the obstacles that his environment presents. In Mirkwood he fares better than the dwarves, because his 'sharp inquisitive eyes' (2008b, 179) allow him to see through the gloom, and further on, when the dwarves have been taken by the spiders, Bilbo manages to find them again by approximating where the cries come from. In a forest that has hitherto been impenetrable and uncooperative, Bilbo's ability to 'guess' where his friends have been taken seems to be more than the 'luck' (2008b, 200) to which the narrator attributes it. Rather, it suggests a specific and developing understanding of the environment he is in; one that upholds the dichotomy between the human and non-human by maintaining the non-human's otherness, its irrationality and danger and desperate need to be tamed.

This episode is particularly striking when considered in comparison with earlier drafts of *The Hobbit*. Crucially, the Mirkwood chapter was the only section of the original manuscript to undergo substantial editing before publication (Rateliff 2011, 335), and one of the greatest changes is how Bilbo navigates the unknown forest. In the original draft, Bilbo – in what John D. Rateliff describes as a take on the Theseus myth – uses a ball of spider thread tied to a tree to trace his steps in the forest and avoid getting lost. In effect, through the use of the spider thread, Bilbo simultaneously maps out his path in Mirkwood and reads the map he has created to guide his way and find his friends. In the final copy of *The Hobbit*, however, Bilbo does not rely on external tools to help him negotiate the unknown forest. Instead, he must read the physical landscape itself to make his way through. The two drafts demonstrate how Tolkien was continually preoccupied with exploring different ways that his characters might interpret and master the environment. Much as Bilbo learns to read the representation of the landscape in Thror's Map, so too does he learn to read, process and draw conclusions from Mirkwood to aid in his own process of identity formation. Unlike the episode with Tom Bombadil, the characters do not engage in a process of understanding the alterity of nature as an emblem of its own subjectivity: Mirkwood and the Lonely Mountain are considered simply dangerous and requiring control and domestication through cartographic instrumentalization.

Maps as objects of knowledge and potential power in unknown and hostile environments continues in *The Lord of the Rings*. Immediately after the Council of Elrond, before the travellers are to set out on their quest, Aragorn and Gandalf – sometimes joined by Frodo – frequently meet together to 'ponder […] the storied and figured maps and books of lore …' (2008a, 360). It is striking that these two characters consult maps in order to increase their knowledge of Middle-earth, as they are arguably the wisest and most experienced members of the Fellowship. Indeed, in his comments on Tolkien's cartographic practices, Tom Shippey argues that Tolkien's characters 'have a strong tendency to talk like maps …' (2005, 100), citing Aragorn and Gandalf as examples of characters that cogitatively and verbally trace and map the land of Middle-earth through their innate knowledge of it. That it is Aragorn and Gandalf who continue to consult maps in preparation for their quest throws into relief their continued unease with the non-human world and their need to maintain hierarchy over it. Just as with the hapless dwarves, meanwhile, the members of the Fellowship who do not bother with maps find themselves lost in the wild. After leaving Rivendell, the Fellowship travels through bad weather and difficult terrain before arriving at the borders of Hollin, where they see the beginnings of the Misty Mountains. Pippin assumes that they had turned eastwards rather than continuing southwards, as they are now facing the mountains, but Gandalf explains that what he sees are the mountains in the distance turning southwest, remarking with characteristic exasperation: 'There are many maps in Elrond's house, but I suppose you never thought to look at them?' (2008a, 368).

The instrumentalization of the non-human through maps finds its other half in an active process of reading and studying the tangible landscape; specifically, Aragorn's reading of maps in Rivendell forms only one aspect of his overall 'land-reading'. A useful term here is Robert Macfarlane's 'literacy of the land' (2016, 23), which he developed in his study of the relationship between Hebrideans and their native home and their use of 'memory maps' rather than paper artefacts (2016, 20). Aragorn's cartographic knowledge similarly translates into an ability to quite literally 'read' the land like a map. Gandalf comments that '[i]f you bring a Ranger with you, it is well to pay attention to him, especially if the Ranger is Aragorn' (2008a, 370), while Gimli emphasizes later on that the weather will not impede Aragorn's ability to track: 'A bent blade is enough for Aragorn to read' (2008e, 636). This language of reading the land occurs multiple other times in relation to Aragorn: when seeking Frodo at the beginning of *The Two Towers*, Aragorn struggles to 'read' Frodo's footprints and has to look closely at the ground several times before he finds the tracks in the earth (2008e, 537). Further on, when searching for Pippin and Merry, Aragorn again 'read[s] the marks' in the ground. Although these are man-made signs to an extent, in that they show human interference in the environment, they are nevertheless embedded in the landscape and inextricable from it: tracks seen in mud and earth, bent and trampled leaves and grass. In Middle-earth, reading maps and reading the landscape form two sides of the same coin: characters need to be either cartographically or environmentally literate (or both) in order to make their way successfully through the world.

Much like cartography, the urgency to read and interpret the signs of the landscape reifies the natural world into a rationalized human framework so that, as Plumwood argues, the non-human only becomes significant and signifying through the supremacy of human reason. Notably, while earlier drafts of the orc hunt in *The Treason of Isengard* depict Aragorn studying the ground and tracking footprints, there is no mention of 'reading' the landscape. As this section of the narrative developed, then, Tolkien began to work in explicitly textual imagery that throws into relief the characters' solipsistic understanding of their non-human environments. The binary between the human and non-human is upheld while – and indeed because – the land is implicated in various human conceptualizations. Absorbed into the anxieties, ideologies and identities of the human, the non-human becomes but a reflection of the human self looking out into the unknown.

For Susan Jeffers, such human/non-human relations in Middle-earth and their political and material consequences can be categorized and understood through a tripartite model. Elves, hobbits and Ents have what Jeffers terms 'power with' nature, which acknowledges the alterity of nature without objectifying it; Dwarves and Men exhibit a dialectic 'power from' relationship, in which the environment is understood through its benefit to themselves; while Sauron, Saruman and the orcs hold 'power over' their environment, predicating their relationship on 'domination and perversion' (2014, 75). Although Jeffers' model accurately unpacks many of the power dynamics in Tolkien's writing, it nevertheless neglects the contradictions and exceptions that we have seen above and that are crucial to Tolkien's broader critique: that anyone (apart from the Ents, as we shall discuss below) is capable of dominating, exploiting and injuring the natural world out of greed and neglect. Aragorn and Bilbo's need to gain knowledge over the non-human does not involve actively harmful interventions onto the environment, but it does betray the desire for 'a measure of control' of which Tolkien was so suspicious. This measure of control may, as Tolkien imagined, lie on a spectrum between the ardently conservative and the wantonly destructive, but the crucial point is that these supposedly opposing attitudes lie on the *same* spectrum: one that sees the non-human world as a resource, obstacle or extension of the human sphere and its loftier concerns. This, then, is anthropocentrism made manifest: unable to surrender to the vitality of the non-human, to the possibility that its presence, purpose, behaviour lies beyond their concerns, even 'good' characters such as Aragorn and Bilbo shore up received understandings of human exceptionalism and entitlement that fix in place extractive hierarchies between the human and non-human.

Stewardship

The supposed exception to this rule, the one model of human/non-human relations that appears – on the surface – to eschew this measure of control and tendency towards self-centredness and solipsism, is that of stewardship. In the introduction to the most expansive study of stewardship in Tolkien's sub-creation, Matthew Dickerson and Jonathan Evans optimistically define the practice as 'the benevolent, selfless custodial care of the environment' as opposed to a '"cover term" justifying the exploitation of our natural resources for commercial, corporate, or personal gain' (2006, xx), and certainly there are examples in Tolkien's legendarium that embody this Platonic ideal. In a model reminiscent of Jeffers' tripartite hierarchies of environmental power, Dickerson and Evans deconstruct the four types of stewardship that exist in Tolkien's legendarium: stewardship as exploitation; an anthropocentric stewardship that continues to centre the human and view nature largely as a resource; a type of authorial stewardship that sees humans both as masters and as managers of the Earth who bear responsibility towards its care (what we might term a conservationist approach); and stewardship as 'servanthood stewardship' (2006, 43), a model that understands the intrinsic value of the non-human world and positions humans as its servants and protectors. This latter model, Dickerson and Evans argue, embodies Tolkien's environmental ethics, an ethics rooted in a particular Christian ideology that understands the Earth as a sacred creation given into the care of the humans who live on it.

Very few inhabitants of Middle-earth, as we have seen, live up to this ideal. The exception, as Dickerson and Evans argue, is Gandalf, whose relationship with Middle-earth explicitly eschews any kind of rule or authority. In a confrontation with Denethor about the defence of Gondor and the protection of Middle-earth, Gandalf tells him:

> But I will say this: the rule of no realm is mine, neither of Gondor nor any other, great or small. But all worthy things that are in peril as the world now stands, those are my care. And for my part, I shall not wholly fail of my task, though Gondor should perish, if anything passes through this night that can still grow fair or bear fruit and flower again in days to come. For I also am a steward. Did you not know?
>
> (2008c, 992)

As Dickerson and Evans note, the nature imagery in this speech is no mere accident. Tolkien explicitly encompasses all life on Middle-earth, from the human to the non-human, as being not only under Gandalf's care, but *worthy* of his care (2006, 44). In doing so, Tolkien depoliticizes Gandalf's stewardship in a very particular way, liberating it from the politics of borders and realms that encourages care only for human-demarcated lands in favour of a more radical environmental politics that centres the non-human within the stewardship of the Earth. The timing of the speech is also very telling. Taking place mere days before Mordor's attack against Minas Tirith and the Battle of the Pelennor Fields, Gandalf's words reorient the significance of the proceedings, refocalizing the purpose of the narrative not merely as a quest or fantasy tale caught up in the tipping point of battles, but as an interrogation of the responsibility and ethics of human behaviour amidst the non-human. Even with the fate of Gondor in the balance, Gandalf understands that its survival is but one aspect of a much broader picture, and the survival of non-human life – things that 'grow' or 'bear fruit and flower' – has as much value as the endurance of a great political and cultural power. This stress on the non-human is repeated later in Gandalf's address to the lords of Gondor and Rohan following the Battle of the Pelennor Fields, in which he urges them to do what they can to 'uproo[t] the evil in the fields ... so that those who live after may have clean earth to till' (2008c, 1150). As Dickerson and Evans remark, even though Gandalf's imperative is not primarily environmental, Tolkien's nature-based lexicon – the uprooting of evil and tilling of the earth – nevertheless centres the well-being of the non-human within the ethical practice of the human. For Gandalf, it is essential, as Dickerson and Evans argue, to exercise stewardship 'in relation to the earth' (2006, 48).

Through Gandalf, Tolkien advocates for a form of stewardship that is inherently ecocentric; the only way to transcend the measure of control that is so inherent to human/non-human relations in Middle-earth is to subsume the interests of the human to that of the natural world. Yet it is striking that, despite this emphasis that Tolkien places on the radical environmental possibilities of stewardship, there are so few other examples of it in his legendarium. Think of stewardship in Middle-earth and the example that likely comes to mind is Denethor, whose rule over Gondor leaves behind a troubled legacy of warped and violent exploitation. It is no accident that Gandalf's reclamation of stewardship and advocation for care of the non-human is made in the halls of Minas Tirith; his rejoinder that 'for I also am a steward' speaks back in juxtaposition to a context of neglect and highly politicized anthropocentrism that defines Denethor's reign. Pippin is startled by the strange internal conflict that seems to rage between the two when they face each other, their eyes near duelling, a strain between them as if 'a line of smouldering fire' (2008c, 990). These wordless interactions, as much as their debates around the future of Gondor, crystallizes the push and pull between their attitudes towards stewardship, emphasizing the utter disparity of their practices: both, as Gandalf says, may be stewards, but the consequences of their positions on Middle-earth's environments could not be more in conflict.

Admittedly, Denethor's exploitation of the non-human environment remains largely symbolic rather than explicit in *The Lord of the Rings*; Tolkien does not depict direct environmental destruction at his hands. Yet the non-human world, encapsulated in the White Tree of Gondor that stands dead in the courtyard, wastes away under his rule. In examining the tension between nature and culture – that is, the non-human and human – in Tolkien's legendarium, Ekman notes the absence of organic nature both in Minas Tirith and in proximity to Denethor (2013, 136); his throne room sits behind the dead tree that is surrounded by a 'sweet fountain' (2008c, 985) and a green lawn – symbols of cultivated and cultured growth that only emphasize the tree's skeletal frame – while his very throne is placed in front of a tree carved from stone and covered in gems as if in mockery of a tree in flower. The non-human is appropriated into human opulence and significations of power or is simply allowed to die; a notable and deliberate monument to neglect in an otherwise, as Pippin notes, 'well tended' place (2008c, 985). The contrast that confuses Pippin, between the dead tree and the cared for city, is no mere accident: it is an inherent ethos of Denethor's rule, where that which can be controlled and tamed – the neat grass of the lawn, the artificiality of the fountain – is allowed to flourish.

Ultimately, as Gandalf himself notes, Denethor is not interested in acting as steward for the various and diverse lands and lives contained in Gondor, but as steward of the idea of Gondor as a ruling power. The White Tree, transformed from an independent, living being into a cyclically dying reminder of the hubris and neglect of Gondor's rulers, becomes but a symbol for the land of Gondor as a political entity. The White Tree is carved in the throne room, at the heart of the realm's power; it is embroidered onto livery, worn by the actants of the land's military might. Denethor's stewardship of the land is inextricably entangled with his political ambitions; in stark contrast to Gandalf, who strives for the survival of human and non-human life in order to fulfil his task of care, Denethor's fears over the impending siege of Minas Tirith centre on governmental rather than material death; his dream is to be steward of the land 'in peace' (2008c, 1118) – that is, power without responsibility, thriving without care. '[M]y line too is ending,' he says, 'the House of the Stewards has failed'. Yet arguably, the House of the Stewards failed long ago, when the White Tree of Minas Tirith became fossilized into nothing more than symbol and synecdoche.

For Ekman, Minas Tirith is saved from this failed stewardship when Aragorn ascends to the throne, reconciling the wilderness of nature with the human construction of the city. He argues that Aragorn's position as a Ranger hitherto in the narrative, as well as his immediate work in the Houses of Healing when he arrives in his city, using the weed-like athelas plant to heal Faramir, Éowyn and Merry, connotes a productive melding together of nature and culture, non-human and human (2013, 139). In opposition to Denethor's flagrant lack of care for the non-human realm within his care, Aragorn's strong connection to the athelas plant suggests a powerful harmony with the non-human that understands the imperative to act as its responsible steward. This is arguably true, yet – as we explored above – Aragorn does not entirely deconstruct his own position of power and instrumentalization over nature. Indeed, looking at Aragorn's rule, Ekman comes to the conclusion that good cities, run the proper way, must mix culture with 'tame' nature, that is, a version of nature that only exists according to human desires and management (2013, 141). This kind of contradiction, or – perhaps more accurately – nuancing of what we understand as 'good' stewardship relationships runs throughout Tolkien's texts. The practice of good stewardship is far more complicated than mere respect for or superficial harmony with nature and there are, in fact, very few ways to do it correctly; even the most positive and well-intended relationships with the non-human, as illustrated by Jeffers' 'power with' dynamics, often betray a problematic attitude to the natural environment.

Flieger delves into this contradiction in her examination of the hobbits' attitudes to the Old Forest surrounding the Shire. Although hobbits are not stewards of their home in the strict political sense that denotes Denethor's position, Tolkien nevertheless stresses their caretaker-like relationship with the Shire: they are fiercely attached to the natural beauty of their home; Sam is a gardener; and their resistance to the industrialization and destruction of nature in the Scouring of the Shire forms the book's final denouement. Yet, as Flieger notes, there is an unsettling discrepancy in this supposedly nature-loving attitude and the behaviour of the hobbits in the Old Forest, when they first set out on their quest. The hobbits encounter their first adversary in the Old Forest, and their brush with Old Man Willow is depicted – from their perspective – as their first run-in with evil on what will become a profoundly evil-laden journey. Old Willow's heart is 'rotten' (2008a, 170) and he is, according to Frodo, a 'beastly tree' (2008a, 154): it is difficult to reconcile this depiction with the otherwise positive presence of trees and tree-ish beings in Middle-earth.

Yet, as Tom Bombadil comes to show them, the Old Forest has reason to be angry and rotten, and it is in dwelling in its interiority – both literally and figuratively – that the hobbits come to understand some of its complexities and perspectives. The result of this shift in perspective – in the recognition of life and experience 'apart from themselves' (2008a, 170) – creates, as Flieger argues, not merely a complication of the Old Forest's morality, but of the hobbits' history and position of power over their land. In particular, the idyllic rural depictions Tolkien makes of the Shire in the Prologue bears an uncomfortable relation to implicit ecological harm. In the very first page of the Prologue, Tolkien describes:

> Hobbits are an unobtrusive but very ancient people, more numerous formerly than they are today; for they love peace and quiet and good tilled earth: a well-ordered and well-farmed countryside was their favourite haunt. They do not and did not understand or like machines more complicated than a forge-bellows, a water-mill, or hand-loom, thought they were skilful with tools.
>
> (2008a, 1)

Here, the narrative ostensibly sets up a contrast between the nature-oriented hobbits and the looming complications of industrialization, positioning the people in their very first introduction as on the side of nature. Yet this pastoral charm and call to parochial simplicity betrays a harmful and destructive environmental practice that is obscured by imagery of countryside bliss. The Shire is 'well-ordered' and 'well-farmed' and filled with 'good tilled earth', yet there is – as Flieger notes – no possibility of good tilled earth without a tree-less land, no potential for farmland amidst self-determined wilderness (2000, 150). Is it any wonder that Old Man Willow's heart has become rotten, that the Old Forest fights back against its own destruction? The hobbits may love their land, yet the history of their actions – hacking and gnawing at the trees of the Old Forest in order to craft their own pastoral idyll – is not that far, in many ways, from the actions of the orcs in Isengard, and the response of the Old Forest is not so very different from the resistance of the Ents and Huorns against Saruman. Indeed, it is interesting that in the film version of *The Lord of the Rings*, Tom's description of the heart of the trees, and their hatred of beings that 'go free upon the earth, gnawing, biting, breaking, hacking, burning: destroyers and usurpers' (2008a, 170), is given to Treebeard: a speech he makes in regards to the orcs on the edges of his own forest. There is an intertextual kind of parallel being drawn here – or perhaps the films simply did not want to complicate this perspective of hobbits, having entirely cut the Tom Bombadil material from the script. After all, as Flieger notes, this kind of thing is 'not comfortable for hobbit lovers to see, let alone acknowledge' (2000, 149).

It may not be comfortable, but, through this kind of tension between narrative heroism and antagonism, Tolkien creates a complicated ecological ethics and standard for stewardship and care that extends beyond mere affection and attachment – on the part of both character and reader. Ultimately, as Sofia Parrlla argues, there is an inherently problematic element to the kind of stewardship that cannot acknowledge the non-human's personhood (2021, 14). It would be incorrect to claim that Tolkien resists an idealization of stewardship in his texts; Aragorn's ascendance to the throne of Gondor is grounded in hope, and Gandalf's selfless stewardship in particular spans across the Ages of Middle-earth. Yet for all his catholic interest in human responsibility over the Earth, it is notable that in the legendarium, successful models of stewardship – few as they are – are distinctly and necessarily ecocentric: there is no mutually beneficial form of stewardship that centres the Anthropos. Parrlla traces this conditionality of stewardship right to the roots of legendarium and the Creation story, where Yavanna, the Vala responsible for the growth and well-being of the non-human, points out to Manwë that many of the trees participated in the creation song of the Ainulindale, urging their protection against the extractive practices of the Elves in the light of this evident mindfulness (2021, 16–18). There is a fine line, Tolkien shows us, between Gandalf and Denethor, care and management, attention and regulation, mutual existence and cartographic control. And being on the right side of history, or the right side of the narrative, is no guarantee of selflessness and true ecological stewardship, no guarantee that the measure of control won't corrupt.

Environmental Destruction

The result of such measures of control – perhaps not as a direct consequence of such indirect acts as Aragorn's land-reading or Denethor's realm-driven rule, but certainly of their mindsets – is wholesale environmental destruction. Although it may seem narratively, or even ideologically, unsound to implicate these more gentle measures within the shocking instances of ecological violence found in the legendarium, it is essential to understand the easy collapsibility between 'power from' and 'power over' relationships, the slippery slope of exploitation that anthropocentrism eventually leads to. We have only to look at the hobbits and the Old Forest, or the Númenóreans, as we shall do shortly, to see how very true this is.

Admittedly, many of the most iconic examples of ecological violence in the legendarium are rooted in a very clear 'power over' relationship, amongst the most notable being the felling of trees enacted by Sauron and Saruman in *The Lord of the Rings*. Strikingly, however, Tolkien demonstrates the vulnerability of the non-human to acts of explicit violence much earlier both in terms of Middle-earth's history and his own textual chronology, through the tale of the Two Trees of Valinor. Originally two lamps set by the Valar to give light to the newly created Arda, the Trees of Valinor are created from the light of the lamps after they are destroyed by Morgoth. In *The Book of Lost Tales*, Palúrien (the early name for the Vala Yavanna) weaves spells of 'life and growth and putting forth of leaf, blossoming and yielding of fruit ...' (1983, 71), all of her energies invested into the nurturing and care of the trees. This is emphasized in a later draft of the 'Quenta Silmarillion', which describes how Yavanna 'hallowed the mould [where the trees are growing] with mighty song, and Nienna watered it with tears ...' (1986, 81); not only does Yavanna create a sacrosanct space around the trees, but Nienna tends to them with her very body, creating an interdependent relationship based on care. Power is markedly decentred in the relationship between the Valar and the trees: Matthew Dickerson and Jonathan Evans draw attention to the language of 'awakening' in *The Silmarillion* – 'thus there awoke in the world the Two Trees of Valinor ...' (J. R. R. Tolkien 2008d, 31) – arguing that this language destabilizes the power inherent in creation myths, instead suggesting that the trees possessed life, subjectivity and material presence prior to Yavanna's song (2006, 8).

Yet this emphasis on nurturing and mutual care is set up only to be undermined by the violence done to the trees by Morgoth: the first explicit act of environmental destruction in the legendarium. Morgoth enlists the help of

Ungoliant – a primeval spirit born of the darkness who takes on the form of a giant spider – to destroy the Two Trees. While the original version in *The Book of Lost Tales* focuses on the loss of light, highlighting the 'fiery radiance' that Ungoliant drains from the tree (J. R. R. Tolkien 1983, 153), subsequent versions dwell instead on the explicitly biomaterial damage. Both the 1930 'Quenta' in *The Shaping of Middle-earth* and the 'Quenta Silmarillion' from 1937 in *The Lost Road* focus on the trees not as bearers of light but as natural matter, with the violence done by Morgoth and Ungoliant framed as ecological ruin: 'With his black sword Morgoth stabbed each tree to its very core, and as their juices spouted forth Ungoliant sucked them up, and poison from her foul lips went into their tissues and withered them, leaf and branch and root ...' (J. R. R. Tolkien 1986, 92). Here, the trees are depicted as organic bodies rather than mere vessels of light, permeable to poison and ecologically vulnerable to the wanton damage caused by Ungoliant in particular, who views the trees as an exploitable source of nourishment and satisfaction. Unlike the original version, the arboreal nature of the Trees is stressed: they wither, 'leaf and branch and root', the accumulative listing emphasizing both the tree's individual and undeniably material existence as well as the idea of total environmental degradation. It is also notable that this episode forms part of Morgoth's theft of the Silmarils and the subsequent First Kinslaying: although Morgoth had certainly introduced evil and conflict into Arda before this, the killing of the Two Trees and the theft of the Silmarils are the events that entirely break the idyllic paradise of Valinor. Not only is the destruction of the trees flanked by two other morally reprehensible acts, but this act of dominance over nature becomes central to the narratives of violence that follow.

The destruction of the Two Trees of Valinor is echoed, amplified and also complicated in the widespread forestry of the Númenóreans as seen in 'Aldarion and Erendis: The Mariner's Wife'; a manifestation of 'power over' nature that breaks Jeffers' racial model and highlights the exploitative potential inherent to all human/non-human dynamics, even among the noble. The tale revolves around the tension between Aldarion, who is drawn to seafaring and voyages Odysseus-like for years at a time to the far-off shores of Middle-earth, and his wife Erendis, who is profoundly attached to the island of Númenor and resents Aldarion's draw to the sea. Although the story is at its heart about the breakdown of a marriage, Aldarion and Erendis' conflict manifests through their opposing attitudes to their natural environments: while Erendis loves nature and in particular forests for themselves (J. R. R. Tolkien 1991, 191), decentring herself within her experience of them, Aldarion's relationship with nature is predicated on the benefits it can provide him. Aldarion's focus on use–value is demonstrated early on: his introduction emphasizes that '[f]rom the first he loved the Sea, and his mind was turned to the craft of shipbuilding ...' (1991, 174), so that Aldarion's affective relationship with the non-human is from the beginning mediated through a pragmatic act that centres the human (that is, himself) within it. This attitude compounds as Aldarion turns seriously to seafaring and desires more and more timber for his ships. His father Meneldur opposes his son's endeavours and forbids him from felling any more trees in Númenor, so Aldarion's sights turn to the shores of Middle-earth, where he 'look[s] with wonder at the great forests' (1991, 176) and establishes a haven to collect timber and build his ships. Aldarion's gaze of wonder might suggest a sublime experience brought about by the magnificence of the forests, yet in actuality, his emotional response is due to their shipbuilding potential: forests as timber, not ecosystems.

At first, however, Aldarion's deforestation seems to be rooted in a relatively moderate and careful practice. There is an emphasis on replanting, both in Númenor and in Middle-earth, so that at times Aldarion gives 'most heed to the future, planting always where there was felling ... new woods set to grow where there was room ...' (1991, 190). Although on the surface this practice of conservation seems to neutralize the damage wrought by the deforestation, it nevertheless continues to trouble Erendis, who suspects that Aldarion still has little care for the forests 'in themselves' and only as potential sources of timber (1991, 191). Erendis' anxieties are central to Tolkien's environmental critique; her Cassandra-like predictions make clear that this hierarchical attitude to nature will lead to inevitable domination and destruction, despite all attempts to the contrary. Aldarion endeavours to replant the trees he fells, yet interlaced with these episodes are others that reinforce his continued power over nature: he is proclaimed 'Master of the Forests' (1991, 181), and although he advocates replanting, there are nevertheless numerous episodes where 'little had been planted to replace what was taken' (1991, 181), and both Aldarion and the narrative become refocused on the felling of trees that are torn apart for the Númenóreans' use (1991, 185). Abandoned both in her marriage and her ecologically minded desires, Erendis bitterly comments that '[a]ll things were made for their [men's] service: hills are for quarries, river to furnish water or to turn wheels, trees for board, women for their body's need, or if fair to adorn their table and hearth' (1991, 207). Her words frame the deforestation not merely as a use of natural resources but rather an exploitation of vulnerable bodies that fits into a broader desire for domination and power.[2]

Power and control are inherently corruptive, assigning the Númenórean deforestation the same thematic significance as the desire for the Silmarils and the One Ring: although Aldarion attempts to practise his mastery over nature responsibly, environmental destruction inevitably reigns. In the Appendix to *Unfinished Tales*, the shipbuilding haven founded by Aldarion in Middle-earth is examined in more detail, the narrative voice switching from the tragic high romance of 'Aldarion and Erendis: The Mariner's Wife' to a more factual yet condemnatory tone. 'Appendix D: The Port of Lond Daer' describes the area around the shipbuilding port as still 'well-wooded', yet emphatically states that its environment used to be different with 'vast and continuous forests' occupying the land (1991, 262).

Here, it becomes clear that Aldarion's gestures towards conservation were insincere and futile. He once again gazes upon the forests with 'wonder' and a 'hunger for timber' that will make Númenor into a great naval force, his willingness to exploit the non-human in order to satiate his desire for power entirely unchecked (1991, 262). Swiftly, the felling of the trees becomes 'ruthless', and there is no longer any thought given to 'husbandry or replanting' (1991, 262).[3] Tolkien exposes the cracks in Aldarion's brand of conservative environmentalism: fundamentally, as Erendis fears, Aldarion does not respect or care for nature in its own right, which leads him to abandon responsible practices as soon as they inconvenience him. That Aldarion participates in and perpetuates a destructive cultural attitude is evident in the way the forestry 'continue[s] to be extended after his days' (1991, 262). As with the Two Trees of Valinor, the tragedy of Aldarion's actions lies both in the violence done to the individual trees and in the broader, large-scale environmental loss for which he is a catalyst. The effects – both of the physical damage done to the environment and the attitude towards nature

---

[2] There is a very evident ecofeminist reading present in Erendis' comparison between Men's exploitation of the natural world and men's exploitation of women's bodies. This is made even more explicit further on, where Erendis warns her daughter Ancalimë about men's selfish ways, pressing her to resist their wills: 'sink your roots into the rock, and face the wind, though it blow away all your leaves' (1991, 207). True to her nature, Erendis uses tree imagery to urge her daughter to strength, thereby reinforcing the connection between the female and the natural body. Although a more in-depth consideration of ecofeminism is beyond the bounds of this book, see Janet Brennan Croft and Leslie A. Donovan's edited collection *Perilous and Fair: Women in the Works and Life of J. R. R. Tolkien* for further discussion of potential feminist readings in Tolkien's legendarium.
[3] Specifically, the felling of the trees is described as 'ruthless' when the native inhabitants of the region realize the deforestation is becoming devastating and begin to resist the Númenóreans' destruction of their habitat. In response, the Númenóreans become more violent, both to the forest and to its inhabitants, and abandon their attempts to replant what they cut down. Here is a very clear link between colonial violence and environmental violence, which will be considered in much greater detail in Chapter 4.

that led to it – are undeniable. The narrative frankly states that '[t]he devastation wrought by the Númenóreans was incalculable' and the tree-felling is described as 'devastating' (1991, 263). This incalculable impact reverberates throughout the rest of the legendarium: during the Council of Elrond, Elrond recalls a time 'when a squirrel could go from tree to tree from what is now the Shire to Dunland west of Isengard' (2008a, 345), an area encompassing the lands around Lond Daer. This imagery of dense forestlands contrasts heavily with the Third Age, where the area – although described as partly well-wooded – is no longer considered sufficiently forested to warrant any depiction of trees on the Middle-earth map, implicating the map in the process of exploitation.

The Númenóreans' deforestation of the area around Lond Daer is emblematic of their broader desire to be in control of their environments and their experience of the world – a desire that will be explored further in subsequent chapters. Although in the case of Lond Daer there is no specific reference to mapping and the kind of control that cartography can offer, the description of the Númenóreans' activities suggests an almost cognitive mapping that takes place. Their interaction with the non-human revolves around two practices: naming and renaming places, and forming and redrawing the layout of the land – as seen with their forestry – acts that are reminiscent of what mapping does on paper. When the Númenóreans first arrive on the shores of Lond Daer they divide the land into two regions split by the river Gwathló, and name one Minhiriath and the other Enedwaith, names that survive and appear on the Middle-earth map from the Third Age. They then begin not only to fell the trees, but to drive 'great tracks and roads into the forests northwards and southwards from the Gwathló' (1991, 262), fundamentally altering the shape of the landscape and quite literally drawing in roads and borders within the country. And not only does this function as a form of mapping in itself, exerting the power that cartography typically provides, it also fundamentally alters the maps that remain. Aldarion's desire for a measure of control, his desperate need to centre himself within the non-human world and his subsummation of the natural environment into his own personal desires finds its legacy in the cartography of Middle-earth: in the non-forested areas and bare tracts of land that concretize and legitimize his carelessly extractive acts.

Non-human Agency

Yet amidst the tree-felling and the mapping, the tracking of ground and the overcoming of mountains, there are other wills at work in Middle-earth. Awesome, indefatigable non-human agency finds its own way through the oppression and destruction of its natural environments, offering an exhilarating resistance to the constant anthropocentrism that attempts to absorb the non-human into its own concerns. These non-human landscapes challenge the domination of the human, whether that be the domestication of nature such as in the Shire or the active destruction of ecospheres such as practised by the Númenóreans.

We see this in the so-called wild environments of Middle-earth, the swathes of forest, mountain and plain that remain untouched, undomesticated and unknown by the human. The division between the human and the non-human is essential in defining the wilderness: Greg Garrard describes it as that which is entirely external to human culture, arguing that wilderness is a relatively recent concept in human history, as it requires the counterpoint of a domestic, agricultural landscape in relation to its own, untamed nature (2004, 59). Notably, then, our definitions of wilderness remain bound up with, or rather delineated against, the human; yet its complete ontological, categorical and material distinction from the human allows it to resist the subsummation to which the rest of the non-human is often subjected, creating space for new, non-human subjectivities and agencies to come to the fore.

Tolkien's depiction of the wild relies on the juxtaposition between domesticated and untamed nature that Garrard points to: his characters' experience of the wilderness in *The Hobbit* and *The Lord of the Rings* feels all the more dangerous when contrasted with the familiar safety they have left behind in the Shire. Many of the texts' most supposedly dangerous episodes – that is to say, those dangerous to the human protagonists – take place in wild spaces untouched by humans: trolls are encountered in the woods, Bilbo and the dwarves are almost eaten alive in Mirkwood, the Old Forest places the hobbits in physical peril on more than one occasion and the mountain pass of Caradhras is treacherous and unforgiving. Notably, it is the very non-humanness of these areas that is dangerous: in Mirkwood, the river will put travellers to sleep, the trees don't allow light or air in, and the spiders that live within will attack and eat passers-by; in the Old Forest, Old Man Willow traps the hobbits and threatens to squeeze Merry in two; and in Caradhras, it is the rocks themselves that fall on the paths, endangering the Fellowship.

The relationship between maps and wilderness is striking. Spaces such as the Shire have their domesticity and concomitant security articulated through precise cartography: although on small-scale maps like the Middle-earth map there is not a noticeable difference between it and other areas, large-scale maps such as the 'A Part of the Shire' highlight its safe quality. The high level of detail and the absence of any warnings on the map give the impression that every part of the Shire is known, safe and can be charted. In his study of fantasy maps, Stefan Ekman comments that the map of the Shire is 'not a map of the unknown, it is very much the known, the labelled, the familiar. It is a landscape tamed ...' (2013, 47), so that the very ability of the fictional cartographer to recognize and piece together the various topographical details, hidden dangers and secrets of the landscape reads as a manifestation of the subjugation of the non-human, a world that does not or cannot resist human control. Yet by contrast, the ultimately unknowable quality of the wilderness resists mapping. As the dwarves remark before they come across the trolls in *The Hobbit*, in certain areas of Middle-earth, '[t]he old maps are no use: things have changed for the worse and the road is unguarded' (J. R. R. Tolkien 2008b, 44), and looking at earlier iterations of this line reveals Tolkien's developing thoughts on the relationship between maps and the environment. In the first edition from 1937, the text originally read: 'the map-makers have not reached this country yet ...' (2003, 69); this was amended to the finalized version in the 1966 reprint. The rewriting is subtle, and on the surface conveys much the same meaning; yet while the original version merely points to the lack of maps as grounds for the alienness of the wilderness, the second version places emphasis on the inability of maps, even when they do exist, to provide a complete sense of safety and control over wild, unknown areas. Moreover, it draws attention to the shift in the maps' effectuality; these areas of land used to be known and calculable, but the increasing danger of the land precipitated by the lack of law and order – revealed in subsequent writing to be due to the failing of Gondor and Arnor as kingdoms of power – allows the wilderness to encroach and frustrates the maps' previous sense of certainty. Other wilderness areas reveal this same resistance: on Thror's Map, Mirkwood is barely featured and therefore untamed: only an arrow points towards the location of the forest, with an ominous note acknowledging the presence of spiders. On A Part of the Shire, meanwhile, the Old Forest is placed on the very margins and spills over the edge of the map, its position recalling medieval modes of map-making, where the unknown would be placed on the peripheries of the known world.

The only map in the legendarium that entirely depicts wilderness is the map of the Wilderland in *The Hobbit*, yet despite the map's attempt to organize, categorize and represent all the dangers in the landscape, it ultimately proves to be useless: it is in this very wilderness, with its focus on the Misty Mountains and Mirkwood, that the company face some of their greatest dangers. Examining the Wilderland map, it reads more as a litany of anxieties about the unpredictable wildness of the landscape than a helpful cartographic document: dragons and spiders populate its breadth, while the hopeful Old Forest Road marked in thick ink through Mirkwood seems almost deluded given the mess the dwarves fall into when trying to cross it. The Wilderland map speaks to the dwarves' remark that maps are 'no use' in these areas – despite their attempt at control and rationalization, the autonomy contained in wild nature refuses the power hierarchy and control that mapping and domineering attitudes to the non-human typically attempt to enforce.

Through such resistance to the map's control, the non-human emerges as something intentional, oppositional and irrevocably alive. This mindful resistance is given extra vigour by the fantastical quality of Tolkien's environments, as he takes advantage of the generic possibilities of his writing to imbue his non-human world with a vibrant agency. His more-than-alive nature not only begins to make a case for the non-human world as its own independent entity, but also – and more importantly – destabilizes some of the fixed categorizations around the human and non-human. It is in these depictions of untameable, chthonic landscapes that we find Tolkien at his most posthuman. Collapsing the boundaries between what we consider the realms of the human and non-human, Tolkien suggests new conceptions of subjectivity, displacing humans as the central actors of the age and, as Haraway imagines, placing the biotic powers of the Earth as the main narrative that can allow for new and radical forms of interdependence. This depiction of the non-human as a deliberate actant is a response to the human/non-human binary's derationalization of nature, yet rather than engaging with notions of rationality (indeed, as discussed above, Tolkien's perception of rationality was entirely divorced from this post-Platonic model), Tolkien closely prefigures Plumwood's arguments by instilling his nature with other, mind-like qualities such as intentionality and emotion. The emphasis on emotion is particularly striking given its historic subsummation to rationality and reason; rather than giving value to these human intellectual hierarchies, Tolkien entirely refigures what it means to be 'mindful' to enact a new type of non-human agency.

We see this in the looming and aggrieved figure of Caradhras, one of the highest peaks of the Misty Mountains that the Fellowship attempt to cross shortly after leaving Rivendell and where they are rebuffed by falling rocks that prevent their continued journey. Notably, the obstacles that the mountain presents are framed throughout the episode as the will of the mountain, rather than mere accident. Gimli comments that 'Caradhras was called the Cruel, and has an ill name' (J. R. R. Tolkien 2008a, 376), giving the mountain a moral and therefore cognizant character; later on, Gimli warns that 'Caradhras has not forgiven us …

He has more snow yet to fling at us ...' (2008a, 379), and argues that the storm and the rock falls are 'the ill will of Caradhras' (2008a, 381), while Boromir adds that 'these stones are aimed at us' (2008a, 376), underscoring the mountain's agency and its capacity to take deliberate actions based on personal feelings and motivations. As it is only a select few (often Gimli) who comment on the ferocity of the mountain in Tolkien's text, this personification of Caradhras could arguably become personal superstition or a rhetorical device intended to underline the physical power of the mountain by the speaker; yet further on in the narrative, it is the narrator who comments, 'with that last stroke the malice of the mountain seemed to be expended, as if Caradhras was satisfied that the invaders had been beaten off...' (2008a, 382). The language used here – forgiven, malice, satisfied – is affective, lending Caradhras the capacity for intentionality and emotional complexity and emphasizing its personal agency. This autonomy is further emphasized by the physical anthropomorphization of the mountain; its 'head' is described as swathed in grey clouds (2008a, 374); 'shrill cries, and wild howls of laughter' emanate from the mountain at the same time as it reacts to the Fellowship and aims rocks at them (2008a, 376); and Gandalf urges the travellers to descend from the mountain's 'knees' (2008a, 383).

Of course, it could be argued that the use of anthropomorphic language replicates the very anthropocentrism that Tolkien is resisting. Yet it is vital to note that the concerns of the human in the portrayal of Caradhras are very much decentred; the use of anthropomorphic language points rather to the reality that the only vocabulary we have to adequately convey non-human sentience and agency is an anthropocentric one. It is, after all, the only framework through which humans have traditionally understood the mind. This tension is addressed by Jeffrey Jerome Cohen in his discussion of American environmental writer Aldo Leopold's phrase, 'thinking like a mountain' (qtd in Cohen 2015, 3). Cohen initially argues that in this phrase, Leopold employs a 'strategic anthropomorphism' that enables greater human sensitivity to ecological independence yet further on, Cohen moves beyond rhetorical devices to consider the ways that such phrases constitute not an anthropomorphization but an acknowledgement of a vital agency that both the human and non-human embody. 'What if,' Cohen asks, 'it is not anthropomorphizing to speak of a stone's ability to resist, its power to attract – and even of its sympathies, alliances, inclinations, and spurs?' (2015, 212). This collapsing of the lexical and phenomenological boundaries between how we imagine human and non-human agency recalls Haraway's centring of the Earth's biotic and abiotic powers, as well as Jane Bennett's theory of 'vibrant matter', in which she argues that everything – the human, the non-human and the inhuman – all possess a 'vital materiality' (2010, 10) that acts upon, amongst and with other forces to produce effects in the world. Similar to Plumwood's call for intentionality, Bennett entirely reformulates the notion of agency so that speaking of nature as active, mindful and intentional is no mere anthropomorphization, but a deliberate configuration of agency and vitality as intrinsic to the non-human. Crucially, Bennett's conceptualization is ontological rather than epistemological; reminiscent of Tolkien's vitalization of his natural world, her approach to agency is not defined by rationality or knowledge, but rather by a thing's subjectivity, existence and experience.

Through its fantastic ability to dislodge rocks and reject trespassers, Caradhras embodies this very same 'vibrant matter', reconfiguring the non-human world as an active agent rather than a passive body and reclaiming the language of anthropomorphization as one that speaks to and encompasses the non-human as well as human. Indeed, despite this supposedly anthropocentric vocabulary, Caradhras firmly constitutes part of the non-human landscape, rather than imagined as a mythological anthropomorphized force, so that the non-human is made literally rather than metaphorically 'vibrant' and active. This is notably not always the case in Tolkien's writing. Verlyn Flieger draws attention to the differences between the Caradhras episode in *The Lord of the Rings* and the scene with the stone-giants in *The Hobbit*, arguing that while the two episodes contain the same markers of falling stone, shrill cries and hostile environment, the former episode is distinctly positioned as the 'intentional activity of the mountain itself ...' (2013, 110). In *The Hobbit*, the stone-giants have much the same effects as Caradhras: they 'hurl [...]' and 'toss [...]' rocks, and their 'guffawing and shouting' can be heard throughout the mountains (2008b, 73). Their presence is very similar to the giants in C. S. Lewis' *The Silver Chair* (1953), whom the protagonist Jill initially confuses for enormous boulders before they begin to move and throw rocks. In both instances, the giants are organically linked to their physical landscape, yet they remain distinct from it. Caradhras, however, is entirely and irrefutably a mountain: a sentient, vindictive mountain, but a mountain nonetheless, so that it is the very land of Middle-earth that becomes imbued with subjectivity and deliberate, vital agency.

And finally, there are the Ents and Huorns, the most famous of Tolkien's human-ish non-humans. Although Ents are a race of beings in much the same way that hobbits, orcs and Elves are, they nevertheless straddle the line between landscape and creature: they are more 'creature-ish' than Caradhras – featuring as they do on the lists and lore of Living Creatures – but, although they are categorized as completely different to trees, their tree-like qualities are nevertheless central to their existence. Tolkien's depiction of the Ents reverses that of Caradhras; while Caradhras is the landscape described anthropomorphically, the Ents are creatures depicted through arboreal language: Treebeard is 'clad in stuff like green and grey bark' which may be his hide; his torso is referred to as his 'trunk' (an appropriately liminal word that can refer to the middle of a human or tree); and his beard is alternately described as 'bushy', 'twiggy' and 'mossy' (2008e, 603), a description that is emblematic of the Ents' ability to slip easily between categories of human and non-human. Treebeard recalls Ents that are growing 'tree-ish', standing still in the forest for seasons at a time with 'the deep grass of the meadow round ... [their] knees' and covered with 'leafy hair' (2008e, 618), all while other trees become more 'Ent-ish'. The Huorns, for example, are very firmly trees, yet have many of the characteristics of a conscious creature: they have a voice, can move, and make intentional and motivated choices.

Through the Ents and the Huorns, Tolkien's examination of non-human possibility reaches its most radical potential. In the midst of human quests and ideological battles, the Ents' presence in *The Lord of the Rings* – and particularly their instrumental role in the Destruction of Isengard – is markedly non-human-centric; it is only when Treebeard connects the felling of the trees with Saruman's increasing power in his environments that he is motivated to take action, displacing the anthropocentric concerns of the Fellowship for a radical recentring of non-human suffering under Sauron's rule, and the biotic powers of the Earth to fight back against it. The Destruction of Isengard, unlike the other sackings that occur throughout the legendarium, is a distinctly non-human overthrow, one that imagines and deploys non-human agency in a way that queries the boundaries around which we define and restrict the human and non-human, and shows the world-changing and life-giving possibility of acknowledging and centring the chthonic vibrancy of the non-human.

Simply put, it is remarkable because it is just so very tree-ish. In regaling Aragorn, Legolas and Gimli with the tale, Merry describes how he and Pippin were borne to the gates of Isengard amidst a veritable wooded army; the air creaking, the night thick with the sound of rustling. The industrial stone and metal of Isengard is no match against their biotic power as Tolkien imagines nature making a wholesale and accelerated reclamation of the exploited and scarred Earth. Merry recalls how the Ents' 'fingers, and their toes, just freeze on to rock; and they tear it up like bread-crust. It was like watching the work of great tree roots in a hundred years, all packed into a few moments ...' (2008e, 739), while later Ents and Huorns stride through Isengard tossing stone 'into the air like leaves' (2008e, 741). The heavy, ominous rock of Isengard is absorbed into the Ents' arboreal powers; roots thread through them and industrial residue is flung leaf-like in the air. 'Stone-cracking' and 'earth-gnawing' (2008e, 742) is how Treebeard himself describes it, as they tear down dams and put paid to furnaces, working to undo the power Saruman, the 'tree-killer' (2008e, 740), has so violently wielded over the non-human.

The non-human as mind-like, the non-human as kinetic, the non-human as angry: through the Destruction of Isengard, the Ents and Huorns defy the lines which have long kept the human and non-human separately defined, and in doing so lay the foundations of a better world for everyone: for the axe-bitten trees and the scorched earth of Isengard and the constricted waters of the River Isen, long kept in check. The culmination of their battle – Isengard filled with sludge, the Huorns clustered around its perimeters – feels like a distinctly posthuman idyll, not only in the unbounded collapse between human and non-human subjectivities and agencies, but in its Haraway-ian portrait of oddkin rather than godkin. Merry and Pippin work to help the Ents and Huorns regain their land, while wood and water and land meld together. The human has been decentred, and through this, a multispecies and cross-ecological flourishing can take place.

The wild agency of these Middle-earth landscapes has just as powerful implications on the cartography of the land. Although there is no specific discussion in the texts about the difficulties of mapping such sentient areas, the general discussions about mapping unknown areas of Middle-earth and Tolkien's chthonic non-human landscape suggest further implicit complications in charting Middle-earth; as difficult as the wilderness and the dangerous areas of the world are to map, it can only become more difficult when what is being mapped is mutable, active and has its own non-human (and anti-human) intentionality. The Old Forest on the borders of the Shire, filled with Huorn-like trees that 'do actually move', has paths which 'shift and change from time to time' thanks to these ambulating trees (2008a, 145), and there are notably no paths marked out on 'A Part of the Shire' through the Old Forest, although other Middle-earth maps do feature roads through forested areas, such as the Old Forest Road marked in Mirkwood in the Map of the Wilderland in *The Hobbit* (although this is itself, as discussed above, barely functional for other, non-human reasons). During the Destruction of Isengard, meanwhile, Merry comments that there was 'a wood full of [Huorns] all round Isengard' (2008e, 740). Tolkien's use of collective noun here is striking; the Huorns are no mere group or cluster but a forest-like assemblage that entirely reforms the familiar landscape. Much as mapping is the ultimate act of non-human instrumentalization, the impossibility of mapping such trees becomes the ultimate act of non-human incompliance: by refusing stillness and passivity, the trees push against the borders, paths and neatly labelled ecospheres imposed by maps, resisting the rationalized, controlling structures that human mapping attempts to impose on the non-human. The personal impetus of the landscape also complicates the concept of an objective map; if the landscape has subjectivity and alters its behaviour depending on who is in it, a single map cannot adequately reflect this to a broad range of readers. Through vital environments such as Caradhras and the Ents, the concept of the 'unknown wilderness' which resists mapping is pushed to the extreme, and the project of mapping as a form of codifying and controlling the natural world is obstructed.

Caradhras and The Destruction of Isengard are two of the most crucial episodes in *The Lord of the Rings*, if not the entire legendarium, because they reform and resist the easy fantasy soundbite of the legendarium – a moral, binary battle between good and evil – into something much more radical and ambivalent: a troubling of our automatic centring of human concerns. Much as Caradhras and the Huorns resist the exploitative presence of humans within their realms, these episodes within the novel resist the solipsism of Anthropocenic thought – both within narrative and cartography – through a paradigm-shifting demonstration of non-human selfhood. Tolkien's undermining of the very maps he himself created, through his reformulation of non-human agency and his fantastic vitalization of the natural world, act as a form of ecological protest that queries the very boundaries of human and non-human distinction. The dwarves are right: once the exploitative human/non-human binary is dismantled, the old maps may be of no use. But through this cartographic disruption, something much more transformative – stranger, wilder and entangled – can emerge.

3

Geology and Time

The fallout of the Anthropocene has, as we have examined, enormous environmental costs; deforestation, ocean acidification and multispecies extinction spread through the planet like a pollution, choking non-human ways of life and possible futures. Yet the Anthropocene is, strictly speaking, a matter of geological rather than ecological fact: environmental damage is a consequence, rather than a deciding factor, in the arrival and acknowledgement of the epoch of the Anthropocene. When we speak of epoch, we are speaking not of recent shifts in climate crisis that have defined the latter half of the twentieth and the subsequent decades of the twenty-first centuries, but of the same immense geological periods that count among their number the Holocene, Cretaceous and Jurassic. The Anthropocene is a stratigraphic concern: the human has become implicated in the non-human rock record, and cannot be shaken out. We are now the greatest deciding factor in our planet's very foundations.

Being geological, the Anthropocene as a concept is deeply concerned with time and the presence of non-human and geological temporalities that ought to exist beyond our scope. We commonly refer to these immense aeonic timescales as 'deep time', a term that Noah Heringman notes speaks to an escalation of scale so profound that time has been displaced into a spatial register in order to be understood (2015, 56). Yet time has always had a spatial component; or perhaps it is more accurate to say that space – and its concomitant material counterparts: land and stone and cartography – has always had by necessity a temporal aspect, existing as it does across time both short and long. Deep time is bound up not only in the history of the land on which we walk but also, as the Anthropocene continues to embed human activity into the resisting realms of the non-human, in the future of our physical planet. Deep time, human time and material space are increasingly becoming entangled, and their different ontological categorizations increasingly difficult to extricate. What does it mean to speak of non-human time, when the human has taken over everything?

In Tolkien's legendarium, time has this same inevitably spatial character, emphasized by Verlyn Flieger in her exploration of Tolkien's preoccupation with temporalities. The round of the year, she notes, encompasses both a period of time and Bilbo and Frodo's respective journeys, while Aragorn positions himself and his companions as 'six days out of Bree' when they arrive at Weathertop, a temporal unit which is in fact a measurement of space (2001, 22). Yet this entanglement between time and space extends beyond such conceptual references into the legendarium's very formulation of timescales, examining both the temporal and spatial precarity that the tension between human and non-human time provokes. This chapter examines the ways in which Tolkien's narrative and cartographical representation of deep time grapples with this modern reformulation of temporal scale that works to disrupt comfortable relationships with time and the place of humans in the world. That the history of Middle-earth is predicated on such enormous and consequential geological change is immediately evident in Tolkien's framing of it as what Humphrey Carpenter has termed a 'mythology for England' (2000, 59); his desire to create a 'vast backcloth' of legends and tales, and his insistence that Middle-earth should not be read as an 'imaginary world' but rather our own set in another, long ago time, immediately directs the reader's attention not only to the formation of his sub-creation but also to its inevitable destruction (J. R. R. Tolkien 2006, 144, 239). As John D. Rateliff puts it, the entire history of Middle-earth is but 'the world's longest line of dominos, set up with infinite care only to be knocked down' (2006, 67), a domino effect that, this chapter contends, is so often geological in nature. Through a combination of uniformitarian and catastrophic events, Tolkien allows the colossal events of deep time to unfold over long timescales and cataclysmic episodes, both representing the dislocating enormity of these scales and also concentrating the anxiety of a changing world, the irretrievable effects of time, and the ecological effects of the human within sudden geological shifts. While the previous chapter confronted the politics of human control over the non-human, this chapter examines how maps become a crucial embodiment of the anxiety of human temporal displacement, becoming a means of desperately trying to fix the human both in place and time.

Deep Time

Tolkien's marked focus on geological change is entangled in a wider preoccupation throughout his legendarium with the passage of time. His dual generic engagement – with quest fantasy and mythopoesis – allows him to play with various temporal scales, from the personal, intimate experience of time to the geological, cosmic span of deep time. This attention to human and non-human temporalities in turn enables an interrogation of cultural relationships with time and, in particular, anxieties that are constructed around time, change and mortality. Although these existential anxieties have been central to the experience of being alive since the beginnings of history, the particular ways in which they manifest are a product of Tolkien's specific cultural and social context, that is to say, of post-Victorian scientific advancements and reconfigurations of temporality. Tolkien may not have been aware of the unintended consequences of the environmental destruction he observed around him that resulted in the Anthropocene's complete upheaval of geological time, but he was well aware of his own contemporary temporal crisis: the discovery of deep time and the dislocation of the human within the Earth's overarching narrative.

Before deep time, the predominant belief, known as catastrophism, theorized that the world was shaped by sudden, cataclysmic events. Catastrophism was largely informed by a desire to reconcile Christian narratives of the Flood and the Apocalypse with geology, and thus theorized a geological framework which incorporated, and was indeed based on, violent catastrophes. Geological time frames of the pre-Enlightenment period were frequently based around biblical evidence: in the seventeenth century, Archbishop James Ussher tried to date Creation, and thus the age of the Earth, by using biblical genealogies, eventually calculating that the date of Creation was 4004 BC; John Lightfoot, vice-chancellor of Cambridge University, refined this to the morning of Sunday, 23 October 4004 BC (Bowler 2003, 4).

Catastrophism, and the belief in what is now called Young Earth theory, persisted well into the nineteenth century. However, in the eighteenth century, uniformitarian theories were beginning to gain traction. Uniformitarianism argued that the world had been formed by gradual geological processes happening over aeonic timescales; the theory was developed and popularized by eighteenth-century geologist James Hutton and nineteenth-century geologist Charles Lyell, who noted that the Earth was in a constant cycle of uplift and creation, erosion and destruction, and that geological change was thus gradual rather than sudden (Gould 1987, 6). Hutton observed two key geological realities: he recognized that granite, as an igneous rock, represented a counter to erosion, as new rock was constantly being created. He also theorized that the breaks in time represented by unconformities found in the Earth's crust – defined as the meeting point of two layers of unconformable, that is to say periodically or materially different, strata – were a result of a combination of erosion and new rock formation (Gould 1987, 6). As the Earth was caught in these constant cycles of uplift and erosion, Hutton theorized, it could hypothetically be millions of years old – laying the foundations for the discovery of deep time.

Lyell supported Hutton's theories in his three-volume *Principles of Geology*, published 1830–1833. Lyell compared the study of history to the study of geology; arguing that much as numerous events, both inconsequential and monumental, shaped the course of the former gradually over many years, the same principle of slow gradual change over time needed to be applied to the latter – a conception of time that began to move slowly away from previous, biblically motivated calculations. With the arrival of the twentieth century came further uniformitarian discoveries that underlined the deep-time theories of Hutton and Lyell and removed religion from the equation entirely: in the early decades of the century it was discovered that certain elements were able to maintain the Earth's central heat thanks to their radioactivity, and thus provide stable conditions for billions of years, once more extending the timeline of the Earth (Bowler 2003, 130). Around the same time, continental drift began to be theorized, which once more reinforced the conceptualization of topographical change occurring over long periods of time, and moved even further from previous religious models.[1]

In uniformitarianism, as in catastrophism, the way the Earth is shifted and moulded and made tangible – its very spatiality – is irrevocably indicative of and tied to how we conceptualize non-human time. The only way to discover and understand the realities of the planet's timescales is to examine the evidence

---

[1] Tolkien's own position on the age of the Earth, particularly as a Catholic, is not as well established. In 1909, Pope Pius X ratified a decree that declared that the legitimacy of the first chapters of Genesis could not be questioned, particularly in regards to the creation of the world and the creation of man. Although the issue of a time frame is never specifically addressed, the decree nevertheless clearly advocated a literal reading. Anne M. Clifford suggests that this strict stance was an aggressive response to Darwin's theories of evolution, which were undermining people's beliefs in the Creation story. However, throughout the eighteenth, nineteenth and twentieth centuries, an approach known as concordism was also practised by many Christians of all denominations. Concordism attempted to reconcile biblical and scientific theories; for example, the eighteenth-century scientist Buffon theorized that the six days of creation were in fact six 'epochs', which would account for the long time frame required by new geological discoveries (Clifford 1991, 221–23). It is not known whether Tolkien subscribed to the church's position or to the more liberal concordist approach. However, there is an interesting parallel between the concordist idea of the days of Creation lasting for epochs and Tolkien's Valian Years, which were the measurement for time before the creation of the sun and the Awakening of Men, and which are much longer than a normal year span. Dimitra Fimi has suggested a broad alignment with the catastrophism model, especially in Tolkien's note about the 'gap' between the end of the Third Age and 'our Days', noting that '[b]y adhering to the older view of prehistory, Tolkien could add 2000 years of history AD to the 4000 years BC, and thus have the end of the Third Age nicely coinciding with the beginning of human history in the Judaeo-Christian tradition … ' (2009, 165).

the physical land offers; in space, we can find all of human and non-human time laid out. This increasing acceptance of deep time was encountered by what David Farrier terms 'a sense of vertigo' (2019, 9) – itself a distinctly spatial metaphor, one of being kinetically out of step with your physical environment. Farrier invokes James Playfair's account of travelling with Hutton to Siccar Point on the east coast of Scotland, where Hutton observed an unconformity in the rocks that he used as irrefutable proof of uniformitarianism. Playfair remarks that 'the mind seemed to grow giddy by looking so far into the abyss of time' (qtd in Farrier 2019, 10), the jagged drop of Siccar Point embodying not only the remnants of a material environment but also the very concept of non-human temporality, the dizzying act of being decentred from the long-duration mechanisms of the non-human world.

Yet this decentring did not last long. A mere century after the discovery of deep, non-human time came the Anthropocene, and with it a reinsertion of the human within timescales we have no business muddling ourselves in. Much like every effect of the Anthropocene, this process was already well underway by Tolkien's time – although he himself would not have been cognizant of it – yet it is useful here to briefly touch on writing that grapples directly with Anthropocenic collapse in timescales, as these theorizations closely mirror Tolkien's own investigation of temporal anxiety and the tension between human and geological time. Heringman describes the effects of the Anthropocene and the offensive of the human within non-human temporalities as an act of 'inscription' (2015, 56), one that seeks to permanently 'writ[e] ourselves into the rock record' (2015, 78). For Heringman, the Anthropocene is an act of leaving legible temporal residue that will be evident millennia hence: the reef gap in the marine fossil record is one such example that will inscribe an irrefutable human marker into the timeline of our planet. Farrier similarly speaks to this idea of encroachment; arguing that one of the most 'unsettling' aspects of the Anthropocene is humanity's intrusion into deep time, Farrier notes a collapse in irreconcilable timescales that goes both ways, our dependence on fossil fuels putting us in intimate touch with a far-distant past while simultaneously – through unprecedented climate change – writing us into the deep future (2019, 6). In both arguments there is an idea of intrusion and interference, an emphasis on the unwelcome, unparalleled and unnatural entanglement of two hitherto alienated temporal scales. It is notable that Heringman speaks of this collapse using a particularly textual language. Strangely recalling the enforced literacy of the land discussed in the previous chapter which cartography (and Middle-earth's inhabitants) attempt to impose, this textual lexicon is a similar act of power: as we write ourselves into the land over time, we write ourselves into time itself. As Farrier argues, past, present and future begin to fold, concertina-like, onto each other, and deep time – once a symbol of a flagrant indifference towards anthropocentrism – becomes but another product of human solipsistic agency. It is perhaps unsurprising, but no less disturbing, that mere decades after we discovered our insignificance in the face of the Earth's timescales, we upended the temporal order entirely and inserted ourselves right back in.

This tension between human and non-human timescales has – much like every entanglement between power and land – its counterpart in cartography. There is a myth of the map as eternal object: the same myth that points to maps as an exercise in rational objectivity pretends that a map can offer, for a long time, an accurate and relevant representation of the non-human world. Yet as critical cartographer Denis Wood argues, a map encodes time just as much as it encodes space; inevitably so, given the profound entanglements between the two. The map, Wood contends, exists within a temporal dimension, and embodies a certain tense: a looking backward or forward towards the landscape it is purportedly representing. In the same way that a map's scale is based on a correlative relationship between the space given in the map and the space of the landscape it represents, so too does a map's temporal scale negotiate the time of the map and the time of world, spanning a particular scale of time and leading to what Wood terms a certain temporal thickness – a hefty stack of time in which the map is continually representative, applicable and exact (1992, 127).

Yet the traditional (that is to say, non-digital) map as object inevitably holds a certain static quality that freezes temporality despite the thickness to which the map pretends. Indeed, it is this very idea of thickness – an attempt at acknowledging temporality – which is perhaps the map's biggest lie. How can a single representation of the land count for a year, two years, ten years, when the non-human landscape never stands still, never waits? In both a catastrophist and uniformitarian framework, in both a realist and fantasy world order, temporal change is inevitable, and temporal change and spatial change are inextricably dependent and concomitant: one very rarely follows without the other. The very act of cartography becomes yet another incursion of human time – and human desire for centrality and control – into non-human temporalities. Bring in the question of deep time and seismic geological change – as Middle-earth and its maps do all too well – and cartography becomes yet another attempted inscription of the Anthropocene into the rock record, one to which Tolkien's enchanted geology is all too resistant.

Middle-earth's Geology

At this point it is worth defining deep time in Tolkien's legendarium, as he never explicitly alludes to the Huttonian concept of deep time as a framework against which he was consciously writing, and the fantastic elements of his work also complicate human and geological temporal scales. Chronologies form an extensive part of Tolkien's world-building, with many of his 'Silmarillion' drafts given in the form of dated annals and texts. In their discussion of creation, the 'Earliest Annals of Valinor' outline that '[t]ime was counted in the world before the Sun and Moon by the Valar according to ages, and a Valian age hath 100 of the years of the Valar, which are each as ten years are now' (1986, 263). The Elder Days that comprise these annals last 3,000 Valian years, or 30,000 in human years. In a draft some twenty years later, Tolkien's timeline and scale shift: the 'Annals of Aman' detail how the Years of the Valar are now:

> longer than nine such years as now are. For there were in each such Year twelve thousand hours. Yet the hours of the Trees were each seven times as long as is one hour of a full-day upon Middle-earth from sun-rise to sun-rise, when light and dark are equally divided. Therefore each Day of the Valar endured for four and eighty of our hours, and each Year for four and eighty thousand: which is as much as three thousand and five hundred of our days, and is somewhat more than are nine and one half of our years (nine and one half and eight hundredths and yet a little).
>
> (1993, 50)

As the Elder Days now stretch to 5,000 Valian years, their actual span in human years is expanded to 47,901. In the Appendices to *The Lord of the Rings*, Tolkien amended this again, terming an Elvish year a *yén* that lasts 144 human years (2008c, 1453), which would expand the Elder Days to 720,000 years. Subsequent Ages of Middle-earth similarly vary according to the stage of Tolkien's world-building, but typically last several thousands of years. In the 'Annals of Aman', meanwhile, there is an explicit blurring of the specificity of timescales: time in Middle-earth is counted from when the Valar come into Arda to create and shape the world, but it is made explicit that previous to this, 'the measurement which the Valar made of the ages of their labours is not known to any of the Children of Ilúvatar' (1993, 50), and that it is only '[a]fter ages of labour beyond knowledge and reckoning' that the Valar begin to count time (1993, 51).[2]

---

[2] See the recent volume *The Nature of Middle-earth* for further discussions on Tolkien's engagement with temporal scale in his world-building.

There are two readings of this temporal framework. First, Tolkien's emphasis on uncounted time 'beyond reckoning' and his expanded timelines of Middle-earth clearly indicate a history that encompasses deep time. However, it could also be argued that, aside from the nod to undocumented time, the actual history of Arda from its creation – that is to say, its geological and environmental formation onwards – is in fact not as long as the billions of years of deep time on our Earth. This is further complicated by the immortal lifespans of certain creatures on Middle-earth, such as the Valar, the Ents, Elves and Tom Bombadil, who disrupt the idea of biological human time as the antithesis to geological time. These important elements of Tolkien's sub-creation make it clear that the critical theories of time discussed above do not simply map onto Middle-earth; however, neither are they irrelevant. Although Tolkien's sub-creation may not have the exact time span of our own deep time, his references to immense years and aeonic timescales nevertheless demonstrates a preoccupation with time outside memory, geological time and expanding temporal scales.

It is useful to touch here on Jeffrey Jerome Cohen's consideration of deep time from a medieval historiographical perspective, which points to the need for a cultural rather than scientific application of deep time. Cohen argues that although medieval people were working within a catastrophist and foreshortened timeline, the enormity of time nevertheless stretched the imagination so that even the improbably compressed Creation story found in Genesis 'has its textual strata, fossils, provocations to dreaming the inhuman, and unexpected geological depths' (2015, 82–83); that is to say, even within a temporally decided refusal of deep time, the machinations of deep time – its immense geological shifts, unimaginable scales and force of non-human agency – find latent narrative expression. In a similar way, the purpose of this chapter is not to argue that Tolkien's sub-creation was mimicking the primary world's exact statistical realities of deep time, but rather to draw attention to the ways his writing was engaging with new anxieties around its discovery and the ways in which contemporary Anthropocene criticism illuminates the tensions between human and geological time that modernity precipitated. There is a particular violence that characterizes human incursion onto deep time which forms part of the broader human/non-human entanglements with which Anthropocene criticism – and Tolkien's fiction – is so concerned.

That Tolkien was engaging with contemporary discoveries of deep time is indisputable through his thematic engagement with time not merely as a temporal or existential marker, but as a geological reality. In the most extensive study of Middle-earth's fictional geology, Gerard Hynes posits that Tolkien shows an awareness of new developments in the geological sciences through his depiction of a gradually changing world, drawing attention to certain passages in the legendarium that are reminiscent of the (at the time) new theory of continental drift. Hynes traces the development of Tolkien's geological ideas from purely catastrophist to uniformitarian: in *The Book of Lost Tales*, he notes that the Vala Ossë drags an island across the sea, pointing out that this most likely derives from an episode in the *Prose Edda*, in which the giantess Gefjon pulls Zealand out into the ocean (2012, 24). The catastrophist underpinnings of this are indicated in a passage shortly before Ossë's endeavour, which explicates '[n]ow this was the manner of the Earth in those days, *nor has it since changed save by the labours of the Valar of old*' (1983, 68, emphasis added), indicating how Tolkien was at the time conceptualizing a static Earth model that only alters through seismic, divine intervention.

However, Hynes locates a shift in Tolkien's geological framework in the 'Ambarkanta' from the 1930s, in which this position is reversed and the geology of Arda is shown to be not just under the influence of the Valar but minute and gradual Earthly forces. The narrative traces the damage done to the world by the various battles of the First Age, and notes too that the physical earth has altered 'in the wearing and passing of many ages' (1986, 240), a uniformitarian principle that emphasizes its gradual pace over a long, sustained period of time. Hynes notes that this is echoed in a later conversation between Gandalf and Glorfindel during the Council of Elrond in *The Lord of the Rings*. Glorfindel suggests disposing of the Ring in the Sea where it might be lost forever (2012, 26), yet Gandalf reminds Glorfindel of the Earth's own capacity for change, remarking: 'seas and lands may change. And it is not our part here to take thought for only a season, or for a few lives of Men, or for a passing age of the world' (2008a, 347). Not only is Gandalf thinking in geological time, highlighting the vast temporal scale of gradual change that occurs outside singular, catastrophist upheavals, his response to Glorfindel also calls to attention the displacement of human concerns and scales within Middle-earth's broader geological movements: the tectonic movement of the Earth is entirely indifferent to their needs and desires. The Ring and the attempts of Men, Elves and Dwarves to be rid of it will not alter the inevitability of geological change, and the Council of Elrond must begin to think beyond human timescales to geological, non-human ones if they hope to destroy the Ring and its symbolic manifestation of exploitative and extractive power.

Hynes' careful tracing of a uniformitarian geological imagination in the legendarium underscores the extent to which Tolkien was considering still relatively recent conceptualizations of deep time in the formation of his sub-creation's history. Notably, the references to uniformitarian change – the narrative of the 'Ambarkanta' and Gandalf's warning about the mutability of the Earth – are both cast in temporal language, framing the spatial shifting of the world in terms of aeonic time and the passing of immense ages. Yet, as Hynes himself notes, the legendarium's engagement with uniformitarianism and its concomitant deep timescales does not preclude catastrophist episodes. These episodes are reminiscent of the religious underpinnings of catastrophism in our own world. Marking moments of monumental and impossibly compressed geological transformation, they are always brought about by supernatural or divine intervention reforming the very fabric of Middle-earth's geology. In the 'Ambarkanta', in response to Morgoth's aggressions, the Valar attempt to create distance between themselves and Middle-earth, so that they 'thrust away Middle-earth at the centre and crowded it eastwards ... and the thrusting aside of the land caused also mountains to appear in four ranges ...' (1986, 239). Here, the effects of the Valar's defence systems very distinctly recall those of continental drift, as new mountains are pushed up and created through the movement of the Earth's surface. Yet the episode is undoubtedly catastrophic; Tolkien explicates a clear causal link between the foundation of new geological features and the action of the Valar, the word 'appear' indicating immediacy, rather than the slow formation of plate tectonics.

The end of the 'Ambarkanta' also touches on two of the fundamental catastrophic events which define the history of Middle-earth and its formation – the destruction of Beleriand at the end of the First Age and the drowning of Númenor during the Second Age – and which complicate the distinctly uniformitarian principles Tolkien was writing around. In both instances, Tolkien describes the world as being 'broken' and 'destroyed', emphasizing the unnatural and damaging nature of these events, rather than as part of a natural geological cycle. In the 'Quenta', the destruction of Beleriand is considered in greater detail:

> Thangorodrim was riven and cast down ... so great was the fury of those adversaries that all the Northern and Western parts of the world were rent and gaping, and the sea roared in in many places; the rivers perished or found new paths, the valleys were upheaved and the hills trod down; and Sirion was no more ... long was it ere [Men] came back over the mountains to where Beleriand once had been.
>
> (1986, 157)

Tolkien emphasizes the sudden and wholesale annihilation of Beleriand through a language of reversal: low valleys are 'upheaved', high hills are 'trod down', and water both disappears from old river beds and appears in new places, so that the world appears to be turned upside down. Tolkien's use of the pluperfect in 'where Beleriand once had been' underscores a sense of finality, and is echoed later in the 'Annals of Beleriand', which end with 'and Beleriand was no more' (1986, 310), framing the radical, seismic and irretrievable shift in Middle-earth's topography: an entire continent immediately disappeared from the Earth's surface. Like the formation of the Northland and Southland mountains, the destruction of Beleriand echoes uniformitarian conceptualizations of geological evolution – the erosive treading down of the high hills, the orogenic upheaval of the low valleys – yet the catastrophic overlay compresses these events within a shortened timescale. While geological events happening in deep time displace the human within Middle-earth, catastrophist events such as these refigure geology back within human timescales. Much like Middle-earth's environment, its very geological foundations are just as vulnerable to external, non-natural forces that disrupt its non-human continuum.

The 'Akallabêth', or the drowning of Númenor, expands on this vulnerability by foregrounding the violent enmeshment between the human and the geological, and the capacity for human action to encroach on geological space and time. A conscious variant on the Atlantis legend and one of the key events in Tolkien's mythology, the 'Akallabêth' details the Númenórean assault on Valinor, spurred by King Ar-Pharazôn's fear of death and his quest for immortality, and the subsequent and entire destruction of Númenor by Ilúvatar in response. This destruction is notably geological in character: the endangered Valar appeal to Ilúvatar, who bends the hitherto flat Earth into a globe, removing the Undying Lands from the physical Earth and drowning Númenor in the process. The 'Akallabêth' represents a key moment both in terms of Middle-earth's world-building and mythology, as it is here that the flat Earth concept which Tolkien originally conceived is changed and replaced by a more scientifically congruent model. Both conceptually and descriptively, the destruction of Númenor is as drastic an event as the destruction of Beleriand, and is notably also catalysed by forces external to natural geological change. The description in 'The Fall of Númenor' is violent: Valinor is described as being 'sundered' from the earth, causing a 'rift' to appear in the sea (1987, 15). Much like the description of the fate of Beleriand, the language here is one of absence, with parts of the old world physically removed. Tolkien also uses the same explicit language of cause and effect, describing how 'Ilúvatar gave power to the Gods, and they bent back the edges of the Middle-earth, and they made it into a globe …' (1987, 16). His repeated use of the active voice, and the invocation of the gods and Ilúvatar as subjects of the action, displaces the geological agency of the world for an external, supernatural agency that is bound up in the actions and consequences of human desire for power.

The influence of these external factors upon Middle-earth's geology and the disruption of deep time scales by human ones offers two potential readings. First, the influence of theological catastrophism is very evident in these episodes: what is striking about the distancing of Middle-earth from Valinor, the destruction of Beleriand and the drowning of Númenor is their unnatural, and indeed their *super*natural, character. While uniformitarian theory ultimately removed God from geology, catastrophism was frequently linked with divine intervention, and Tolkien closely models Middle-earth's geology on this characteristic. Each cataclysmic shift in Middle-earth's history is the result of the gods' actions, whether indirectly, such as the Great Battle where Beleriand is destroyed through a large-scale conflict between the deities, or directly, such as in the 'Ambarkanta' or 'The Fall of Númenor', where the Valar and Ilúvatar are instrumental in repositioning and restructuring the world and warping the natural flow of deep time in favour of immediate change. Their domination negates lithic and geological agency, refusing the independent mutability of rock and the alienation of the human from both geological activity and its temporal scale, and instead reprioritizing human frameworks of signification, such as the divine, the mythological and the interventionist.

However, the destruction of Númenor can also be read even more emphatically as a signifier of human encroachment into non-human space and time. Its cataclysmic event is notable for its anthropocentric roots: the world is globed and Númenor is destroyed because the Númenóreans demand immortality and the Valar seek to defend themselves. More than just a fantastic interpretation of the Atlantis myth and an existential tipping point for Númenórean descendants such as Aragorn, the drowning of Númenor has a very distinct environmental and topographical aftermath: the destruction caused by this globing takes the concept of human-induced environmental damage to the extreme, by having human activity affect the very foundation of the world. A later account of the destruction details:

> for in some places the sea rode in upon the land, and in others it piled up new coasts. Thus while Lindon suffered great loss, the Bay of Belfalas was much filled at the east and south, so that Pelargir which had been only a few miles from the sea was left far inland, and Anduin carved a new path by many mouths to the Bay. But the Isle of Tolfalas was almost destroyed, and was left at last like a barren and lonely mountain in the water not far from the issue of the River.
>
> (1996, 183)

The tidal aftermath of Númenor's destruction entirely reconfigures and reshapes Middle-earth's coastline, once more compressing gradual activities of erosion and sedimentation into a preternaturally shortened timeline. It is difficult, meanwhile – particularly as a modern reader – not to read the account of the flooding, its loss and destruction and the remaining barren, stripped landscape as an eerie premonition of climate change's looming deluges. Much like in our own age of the Anthropocene, the non-human is made complicit in its own destruction: the force of the sea gouges and flattens the land, doing in mere days what the Earth's natural geological cycle ought to do in the wearing of innumerable years.

The intrusion of the human into the geological in this episode remarkably prefigures the damage of the Anthropocene era, and the enmeshment of the two temporal scales. Both the destruction of Beleriand and Númenor are dense with explicit and implicit loss and death: entire lands and their people pulled beneath the surface of the sea in a way that recalls the Anthropocene's own precipitation of death and destruction. Evidently Tolkien would not have known of the specific geological, ecological and biodiverse impacts of the Anthropocene, such as plastic pollution, ocean acidification and nuclear radiation, that would characterize the enmeshment between the human and geological; yet it is striking, and almost eerie, to see how he expanded his wariness of human interference in the natural world to include large-scale, geological shifts in a way that speaks to his future and our current ecological concerns. Each of Middle-earth's episodes of geological change is striking for compressing long-scale, deep-time uniformitarian change into a human conceptualization of temporality in a way that mimics the fundamentals of catastrophism that long defined our understanding of time; yet there is also an expanding of human time into the geological in a way that preludes the harmful intrusion of Anthropocenic change. The Númenóreans' desire for power, and specifically for an immortality that would supersede their human temporalities and place them firmly within non-human time – as we shall discuss further below – catalyses specific foundational changes that remain inscribed in Middle-earth's geology for aeons after the Númenóreans themselves have passed on. They have, as Heringman phrased it, written themselves into the rock record in ways that cannot be undone, and in doing so have collapsed the distinctions that exist between human and deep time. The history of the Númenóreans' violence and their desire for control over their own brief and fragile temporality overlays the present make-up of the Earth's topography and geology, and will continue to make itself felt in the future.

Tolkien was profoundly preoccupied by what he called his 'Atlantis-haunting', describing a recurrent dream of an 'ineluctable Wave, either coming out of the quiet sea, or coming in towering over the green inlands …' (2006, 347): the drowning of Númenor long before he wrote it. He transferred this persistent vision of destruction to Faramir, whose dreams of the drowning of Númenor became not a premonition of a story, but a lingering reminder of ancestral hubris and its unfathomable consequences on the world, a wave that climbs over 'green lands' and a 'darkness inescapable' (2008c, 1261) that blots this flourishing of the land. His words find uncanny resonance in Anna Tsing's examination of the Anthropocene, which she describes as a kind of haunting: imagining landscapes that are composed of assemblages both past and present, some naturally deceased and others filled with the ghosts of human-precipitated ecological destruction (2017, 2). Nils Bubandt similarly draws attention to the enmeshment of death and our ecological crisis; pulling the Anthropocene into the realm of Achille Mbembe's necropolitics, he argues that all human and non-human life now lives and dies according to human behaviour, bringing geology and its concomitant temporalities under the purview of human politics (2017, 125).

The existential haunting that Númenor signifies for both writer and character has its counterpart in the kind of geological, Anthropocenic haunting that Tsing and Bubandt lay out. The Númenóreans' desire to – as Anna Vaninskaya frames it – 'stav[e] off' time (2019, 168) or – put another way – to ensure their presence within non-human time ironically condemns both themselves and the environment around them to death, to the very finite structures of human time which they were trying to escape; yet the reverberations of their betrayal is felt across Middle-earth's ages. The remnants of Númenórean violence are writ large in the abruptly and irreparably changed structure of the Earth; human time is brought harmfully into non-human temporalities, and residues of death are inscribed in and on the surface of the Earth, in the 'great loss' and 'destroyed' topographies of the land (J. R. R. Tolkien 1996, 163). Two tales which expand on the haunting, latent presence of Númenor throughout the legendarium, meanwhile, are Tolkien's two incomplete time travel tales, 'The Lost Road' and 'The Notion Club Papers'. In both, movements through time become inextricably tied to the moment of Númenor's collapse, and the moment of collapse is inscribed across temporal strata. In 'The Lost Road', the scholar Alboin is visited by strange languages and dreams that have come to him since his childhood, remnants of Eressëan and Belerendiac that are pulled from some latent, subconscious state. Twice, Alboin looks out to sea and says aloud of the clouds, without knowing why: 'They look like the eagles of the Lord of the West coming upon Númenor' (1987, 38), a reference to the Eagles of Manwë that appear in every account of the 'Akallabêth' and act as a portent of the coming storm and flood that precipitates the island's demise. This same phrase is uttered in 'The Notion Club Papers' by Alwin Arundel Lowdham who, as if possessed, speaks in a 'changed' and 'ominous' voice: 'Behold the Eagles of the Lords of the West! They are coming over Númenor!' (1992, 231). Both Alboin and Lowdham speak automatically, compulsively, as if are barely conscious of saying it; Lowdham in particular, whose voice is entirely changed, does not seem to speak the words as his own conscious, intentional self. There is an eerie, supernatural character to their inexplicably inherited memories that speaks to the haunted quality that the drowning of Númenor holds over the entire legendarium, its necropolitical residues of change and violence stretching unnaturally over time and consciousness.

It is particularly striking that both Alboin and Lowdham's haunted recollections point explicitly to the moment of Númenor's destruction – the great, eagle-like clouds that bring lightning and obliteration to the land – rather than any other moment within Númenor's, or indeed Middle-earth's, history. The very tipping point of human temporal intrusion into non-human time stretches disruptively across timescales, to Alboin's time in the early twentieth century, and to Lowdham's in the (for Tolkien) futuristic 1980s. Christopher Tolkien's commentary on 'The Lost Road', meanwhile, reveals the extent to which these temporal reverberations would be felt, foregrounding Tolkien's intention that 'in each tale a man should utter the words about the Eagles of the Lord of the West, but only at the end would it be discovered what they meant' (1987, 77), so that the physical portent of Númenor's destruction is accorded more narrative weight than its psychological effect on the characters. A series of phrases that had 'come through' (1987, 46) to Alboin one morning similarly dwell on the images of annihilation and upheaval that signify the 'Akallabêth'. 'Ullier kilyanna ... nūmenōre ataltane' becomes 'poured in-Chasm ... Númenor down-fell' (1987, 47), a quasi-Modernist poem of fragmented apocalypse that emphasizes the geological and ecological nature of the event: poured in waters and rent apart tectonic plates and an entire landmass falling into a chasm.

It is in the destruction of Númenor and its continued ripples across Middle-earth and our own world's epochs that Flieger's comment on the enmeshment between time and space in Tolkien most comes to bear. The tale embodies a state of temporal, and thus by necessity spatial, intrusion that on occasion discomforts even the characters themselves. In 'The Notion Club Papers', scholar and friend of Lowdham's Michael Ramer experiences similar lucid dreams of the destruction of Númenor and 'a Green Wave, whitecrested, fluted and scallop-shaped' (1992, 194) towering above trees and land, as well as other dreams that manifest a kind of time travel. In one instant, he experiences latent memories from a meteorite years after happening across it in the fictional Gunthorpe Park in Matfield. He works back through the meteorite's history in a moment that very much recalls the Fellowship's encounter with Caradhras and Tolkien's exploration of lithic memory and agency:

> But above, or between, or perhaps through all the rest, I knew endlessness. That's perhaps emotional and inaccurate. I mean Length with a capital L, applied to Time; unendurable length to mortal flesh. In that kind of dream you can know about the feeling of aeons of constricted waiting ... I found it all very disturbing ...
>
> (1992, 182–83)

Ramer interlopes briefly in non-mortal Long Time – a profoundly poetic premonition of the Anthropocene's conceptualization of deep time – and finds himself unwelcome. He understands that it is 'unendurable' precisely because he has no place in it: this type of time was not created for 'mortal flesh'. Yet Ramer can leave his dream and pull himself out of this collapse of temporal boundaries between human and non-human time. The rest of Middle-earth is not so lucky. As the destruction of Númenor illustrates all too well, the potential hostility and discomfort contained in the Earth's endlessness is no protection against human intrusion, and – when it comes – human temporality has the ability to entirely upend the even the vastness of geological time. After all, what better way to refuse the indifferent flow of non-human time than to insert yourself directly into it?

Mapping Geology and Geologizing Maps

Much as maps encode and represent the physical landscape of the non-human world in order to manifest control over it, as we saw in the previous chapter, so too do they attempt to exert this same control over geological spatiality and non-human temporality by bringing non-human temporalities within the scale of the human and fixing land – and time – into place. Tolkien's small-scale maps – that is to say, the Middle-earth map, the map of Beleriand, the 'Ambarkanta' maps and the map of Númenor, all of which are best able to represent large areas of the Earth – pretend at a concept of temporal thickness that acknowledges temporalities beyond the human yet, as with any map, this thickness is entirely negated by the inherent flattening of cartography into human time.

That is not to say that Middle-earth's cartography entirely refuses to encode non-human temporalities. Most obviously, each of these maps works to visualize the physical changes described in the text; a visualization that works particularly on the level of an imaginary or extradiegetic reader who can immediately cross-reference the maps with a knowledge of Middle-earth's history or the texts detailing the shaping of Middle-earth. Particular paratextual details on the maps, meanwhile, indicate that the diegetic Middle-earth map-maker and map-reader is aware of and attempting to record the enormous shifts that characterize Middle-earth's geological formation. The third diagram of the 'Ambarkanta' maps bears the inscription, 'The World after the Cataclysm and the ruin of the Númenóreans'; similarly, the revised Middle-earth map found in *Unfinished Tales* is labelled 'The West of Middle-earth at the End of the Third Age'. Both of these cartographic paratexts embed not only time into the map but also a sense of time passing: the 'Ambarkanta' map points to the idea of a before and after, and a need for remapping, while the Middle-earth map, while less explicitly cataclysmic, situates the map very firmly at the closing of an era, accentuating, as Stefan Ekman argues, not just a sizable past of three ages but also 'a future, a Fourth Age from which it is possible to establish the end of the previous age' (2013, 61).

Another more explicit visualization of the enormous shifts in land appears in the Middle-earth map when read extradiegetically alongside the map of Beleriand, which illustrates Middle-earth during the First Age. The map of Beleriand depicts a range of mountains named the Ered Luin, which wind down the entire eastern border of the map in a straight line from north to south. These mountains are the only topographical element from the map of Beleriand to also be depicted on the Middle-earth map. On the Middle-earth map, however, they are depicted in the extreme west, and their long, unhalting line across the longitude of the continent is now disrupted: they curve along the coast in the north-western corner of Middle-earth, and are then interrupted by the Gulf of Lhûn, before resuming again as a short range named the Blue Mountains. To the west of the mountains, just off the coast, is a small island named Himling, a name linguistically very like the mountain of Himring that appears on the map of Beleriand to the west of Ered Luin, just past the March of Maedhros. Christopher Tolkien confirms in *Unfinished Tales* that 'Himling was the earlier form of Himring [...] it is clear that Himring's top rose above the waters that covered drowned Beleriand' (1991, 13–14), recalling what also happened to the Isle of Tolfalas, the barren and lonely mountain left after the drowning of Númenor. In a world-building framework where the landscape is constantly undergoing physical upheaval and nomenclatures constantly change, it is striking that traces of the Ered Luin mountains and Himring remain, that their names remain the same or recognizably similar, and that they are depicted on both maps. Tolkien visualizes their presence in order to suggest the absence – and haunting – of a former geological reality, implicating what has been destroyed by showing what remains.

Here, the maps are attempting to encode a type of temporality, yet it is probably more accurate to say that they are depicting the temporal tension that exists between human and non-human timescales. The maps gesture towards time, yet they – by their textual nature – depict a very fixed version of Middle-earth that cannot encompass the changes, slow and quick, that take place in the Earth's geology. Instead, they act both as another example of pulling the geological within the human, freezing the deep time of the landscape into human representation, and as proof of their own obsolescence – of the impossibility of inscribing geological time using human tools. These cartographic artefacts endeavour to exercise control over the land yet are to a large extent thwarted by the non-human character of the land that exists beyond the temporal thickness to which the map can aspire. We see this in *Unfinished Tales*' 'A Description of the Island of Númenor', a melancholy elegy for Númenor's disappeared history, which explains that what is still known of the island is taken from accounts and map preserved in the archives of Gondor (1991, 165). Yet with Númenor entirely gone, these maps can no longer serve their original, intended function as navigational tools and have become obsolete as cartographic tools. Instead, they have turned into historical artefacts – objects which can offer a window into the past, and which can speak to the catastrophic nature of the world's geology, but which are otherwise redundant. Their placement within the archives of Gondor further underscores their inadequacy as cartographic objects. Unlike other examples of maps in Tolkien's legendarium, such as Thror's Map in *The Hobbit* which is carried around by the characters and is constantly in use, the maps of Númenor have become static, fossilized as records of history rather than geography, rendered useless by the impossibility of the map to record non-human temporalities. Their stultified position in the locked archives of Gondor reflects what Vaninskaya broadly refers to as the Númenóreans' 'mortuary culture' (2019, 187), a state of existence dominated by extinct relics of the past so that the Númenóreans 'built mightier houses for their dead than for their living' and the 'west shores of the Old World became a place of tombs, and filled with ghosts' (J. R. R. Tolkien 1987, 16–17). The maps become another aspect of this tomb culture, of a land and people more preoccupied with death and the insufficiency of human time than with reconciling their diminutive place within the unfathomable scale of deep time, crafting fossils and regret from their past.

This comparison between maps and fossils is a deliberate one: it speaks germanely to this idea of maps as obsolete relics, as well as to the uncomfortable Anthropocenic haunting of Middle-earth's past landscapes throughout the texts. Yet this analogy is not merely intended as a metaphoric device. Rather, it highlights a particular function of maps within Tolkien's legendarium to provide the kind of geological residue that fossils would normally provide, a crucial ontological deposit that was being increasingly understood by the scientific community immediately before and during Tolkien's own time as a divisible marker between human and non-human time. By the end of the eighteenth century, fossils had become curios and objects of both scientific and commercial interest: areas of England where they were abundant, such as Dorset and in particular Lyme Regis, became tourist attractions, and residents of these areas frequently collected fossils from the beaches to sell to tourists (Cadbury 2012, 6). Not everyone appreciated the scientific and historical implications of fossils, with many turning to a supernatural explanation: fossils were people turned into snakes for their crimes, petrified thunderbolts from God or the material spirits of animals (Cadbury 2012, 7–8). However, naturalists and scientists began to recognize them for what they were: remnants of a geological past. In particular, French naturalist and zoologist Georges Cuvier worked on classifying fossils, extrapolating from fossilized remnants what the original animal might have looked like. Cuvier theorized the idea of extinction, arguing that fossils were an undeniable proof of animals that no longer existed in the nineteenth century and this acted as a record for what was no longer there.

By Tolkien's time, fossils and palaeontology had become a part of the public consciousness and popular culture: the Natural History Museum in London opened in 1881; the Natural History Museum in Oxford had opened some decades previously in 1850; and numerous literary and artistic works in the late nineteenth and early twentieth centuries, from E. Nesbit's *Five Children and It* to Noel Streatfeild's *Ballet Shoes*, played with ideas of palaeontology. Tolkien himself had at the very least a passing childhood interest in fossils. He holidayed four times in Lyme Regis, twice as a child and twice as an adult; on his second childhood trip, Hammond and Scull note that he searched for fossils in the cliffs and found a prehistoric jawbone (2006, 12). It is notable, then, and slightly strange, that in spite of the attention he gives to his sub-creation's geological history and its long temporal scales, there are no mentions of fossils in Middle-earth.

This absence could be explained by Tolkien's positioning of Middle-earth as the prehistory of our own time; thus, rather than fossils, he creates a world which contains the living counterparts of our modern fossils. Indeed, many of Middle-earth's creatures are given a distinctly prehistoric characterization that frequently map directly onto species from our own world's prehistory. Of the mammoth-like Oliphants, Tolkien writes: 'the Mûmak of Harad was indeed a beast of vast bulk, and the like of him does not walk now in Middle-earth; his kin that live still in latter days are but memories of his girth and majesty …' (2008e, 864). Middle-earth's dragons, while having distinctly mythological roots, also bear certain similarities to dinosaurs within this prehistoric context, drawing perhaps on the Victorian practice of referring to dinosaurs as dragons; in a 1938 lecture at the University of Oxford Museum, Tolkien begins his discussion on dragons with an overview of prehistoric reptiles (2018, 39). The fellbeasts of the Nazgûl, meanwhile, are deliberately pterodactyl-like: they are described as looking 'like bats' (2008e, 774) and having 'bird-like forms, horrible as carrion-fowls yet greater than eagles' (2008c, 1058).

Yet, crucially, the most detailed description of them in 'The Battle of Pelennor Fields' suggests that these creatures are considered prehistoric even in Middle-earth's time:

> It was a winged creature: if bird, then greater than all other birds, and it was naked, and neither quill nor feather did it bear, and its vast pinions were as webs of hide between horned fingers … a creature of an older world maybe it was, whose kind, lingering in forgotten mountains cold beneath the Moon, outstayed their day, and in hideous eyrie bred this last untimely brood …
>
> (2008c, 1099)

Tolkien reiterates that these are creatures 'of an older world' in a letter to Rhona Beare in 1958, where he explains he did not intend the fellbeasts to be pterodactyls, but that they are nevertheless 'pterodactylic … [their] description even provides a way in which [they] could be a last survivor of *older geological eras*' (2006, 282, emphasis added). The fellbeasts, as well as animals such as dragons which are ancient beings that are all but extinct by the Third Age of Middle-earth, make evident that there is a prehistory that predates even Middle-earth's own relatively prehistoric time frame. In early notes on 'The Lost Road', meanwhile – when Tolkien was considering how to bridge the story of Númenor with our contemporary timeline – he sketches out ideas for where and when

Alboin could visit, including 'painted caves', 'the Ice Age – great figures in ice' and 'before the Ice Age: the Galdor story' (1987, 77), while a later chapter outline includes Chapter V 'Prehistoric North: old kings found buried in the ice' (1987, 78); all suggestions of historical and indeed prehistorical moments that the time-travelling hero could visit that predate the Middle-earth from his legendarium. Although 'The Lost Road' never came to fruition as a Middle-earth story, Tolkien's consideration of prehistoric elements and in particular of preserved remnants as part of the main narrative indicates an intention to construct and iterate a prehistoric past and a particular kind of prehistoric residue.

Despite the creation of this prehistoric background and his imagination of ancient creatures, then, Tolkien's final published version of Middle-earth includes no encounters with fossils. In *The Hobbit* and *The Lord of the Rings*, the ancient character of the world is frequently discussed, but it has no physical manifestation in the form of palaeontological remnants. Instead, the archival role that fossils would play, where they act as a record of the Earth's physical history – as Cuvier successfully discovered – is fulfilled by objects such as the Middle-earth maps, fixed relics of a bygone era. I do not mean to argue that the parallel between maps and fossils was a deliberate metaphor, or that the substitution of the former for the latter was consciously done; rather, this comparison is intended to demonstrate how Tolkien turned to other devices to provide the same sort of historical residue that fossils were providing elsewhere in contemporary science and popular culture. Much like fossils, Tolkien's maps are a material remnant of past geological eras, conserved for their historical value and acting as insights into the previous state of the world. They can speak to the age of the world, reifying the time span between what they depict and their present condition, and providing tangible proof of this passage of time. Fossils are the relics of what was once alive and active; maps, too, can offer frozen examples of a much more complex and evolving world.

Yet maps, ultimately, are not fossils, and their role as provider of historical residue is as much an indication of human intervention into the non-human as it is a representation of deep time. Fossils are a way of bringing deep time into view, bringing its unfathomable and non-human reality into the material present. When maps become fossils, they try to establish a new kind of material past, one that attempts to undo the passage of deep time by fixing in place the human experience of the geological past and one that by its very nature fails, embedding into historical artefact Vaninskaya's concept of 'mortuary culture' rather than a portrait of human vitality and control over time's inexorable movement. The static nature of the Númenor maps in the Gondorian vaults – both in terms of making static the landscape they depict and their own static position within the archive – lends them many of the connotations of fossils: inert, dead and extinct. They make evident not only the extreme gap between what they represent (the landscape of the past) and what they are now but also the inability to bridge this gap: manifesting and visualizing the tension between human and geological temporalities and the futile attempts of the former to fix and control the latter. Much as Tolkien's environment, despite the marked destruction it suffers at the hands of the humans of Middle-earth, employs non-human agency and power to resist human control (both material and cartographic), so too do Middle-earth's geological and temporal timescales resist the incursion and collapse of the human into their realm. The Númenor maps are an exercise in nostalgia, a crystallization of a better time that is now entirely useless: everything depicted on the Númenor map in *Unfinished Tales*, from its mountain ranges to the carefully drawn dotted roads – the only remnants of a vanished civilization – is now gone. Despite its attempt at self-awareness, The West of Middle-earth at the End of the Third Age will fast become irrelevant: the Fourth Age will come, and then the next, and then the next, and what the map depicts will vanish entirely. Himring has already vanished and Himling will soon follow, and with it these cartographic attempts to establish human temporalities in Middle-earth's deep history. The line of dominos will continue to fall and the old maps will, once more, be of no use.

Fixing Experiences of Time

These cartographic exercises over the inexorable movement of time speak to a particular cultural relationship with temporality, for deep time – as we have seen – has not been permitted to remain a non-human, geological fact, but has transformed into an existential state for the humans living ever so briefly within it. As Flieger notes, the theme of time scaffolds Tolkien's entire oeuvre; from the fiction of his legendarium to his critical output, his writing reveals a profound preoccupation with time passing, the (im)possibility of reconciling oneself to its indifferent thrust forward, and the lingering ghosts of the past found in memories, dreams and painful nostalgia. Tolkien's exploration of the Old English poem *Beowulf* in his seminal lecture and subsequent essay 'The Monsters and the Critics' could well be written about his own legendarium: no matter the structures of heroism and tragedy that dictate the narrative of the poem, Beowulf will always lose, for 'the contest on the fields of Time' can result in nothing but a 'great temporal defeat' (1997b, 22). Turning his eye to his Middle-earth writings, Tolkien echoed these same ideas in a letter to Joanna de Bortadano, stating plainly: 'The real theme for me is about something much more permanent and difficult: Death and Immortality: the mystery of the love of the world in the hearts of a race "doomed" to leave and seemingly lose it; the anguish in the hearts of a race "doomed" not to leave it, until its whole evil-aroused story is complete' (2006, 246). Death, of course, is the defining characteristic of human temporality, bookending its flow with a decided finality with which Tolkien's mortal characters wrestle to come to terms. The theme of time, and Middle-earth inhabitants' cultural and political relationship with time, is inextricable from the spectre of death that hangs over all the races.

'The Fall of Númenor', of course, crystallizes this struggle with the finality of death and the yearning towards non-human time, bound up as it is in geological temporalities that throw into relief not just the tension between human and non-human timescales but the fear of death and finality that precipitates this tension. It is useful here to return to Vaninskaya's exploration of what she terms the Númenórean mortuary culture, a hegemonic state of social and political systems informed by and constructed from the Númenóreans' relentless fear of death. This death cult, Vaninskaya argues, existed long before the cataclysmic loss of their island and their subsequent exile to the shores of Middle-earth where their long lifespans began to increasingly shrink. Rather, the Númenóreans have long 'murmured against [the] decree' and natural order that condemns them to mortality, seeking 'unceasingly' for secrets that might extend their lives and turning against the very gods who gave them their already more-than-human lifespans (1987, 15). The geological destruction of Númenor is catalysed by their desire to bring themselves into non-human time, but specifically is a result of their anxieties around death and the smallness of their timescales in the face of much bigger ones.

These anxieties mimic the 'Age of Anxiety' in which Tolkien lived and out of which he wrote, an age that Flieger contends is the underlying factor in his prevailing thematic preoccupation with time, and which is entirely informed by the sudden confusion, displacement and wreckage of burgeoning modernity (2001, 7). In exploring the evidences of this 'Age of Anxiety' in the legendarium, Flieger dwells on specifically political examples that speak to the 'disassociation, dislocation, and psychological ravagement of modern life' (2001, 2): the dystopian hellscape of the Scouring of the Shire, the increased isolation of the Hobbits from a world that moves too fast. Her examples indicate an unquestionably pertinent yet typical conceptualization of the psychological weight of modernity that is rooted in Victorian and post-Victorian advances in technology and the aftereffects of the World Wars, a state of affairs that left behind an unrecognizably violent, fast-moving world. Yet these anxieties around modernity are also, I would add, very much rooted in developing understandings of time itself, and the dislocation of the self within large timescales: it is this that we see reflected in the moral dissolution of the Númenóreans. Their anxieties around death are very much a response to the new possibilities of infinity and immortality that deep time suggested, so that even after the punishment of the drowning of Númenor, 'desire of long life upon earth' sits as heavy on them as 'the thought of Death' (1987, 16). In a later version of the 'Akallabêth', titled 'The Drowning of Anadûnê', we – and the people of Númenor – are told that '[t]he Valar may not withdraw the gift of Death' (1992, 333), yet in the face of long life, death for the Númenóreans is nothing more than a curse. Indeed, in a grotesque recollection of the entombed maps in the Gondorian archives, the Númenóreans learn only the art of 'preserving uncorrupt for many ages the dead flesh of men' (1987, 16), a fossilization and embalming that is but a faint mockery of the immortality for which they long.

The fossilization of maps and the world around them is an expression of this death cult and desire for temporal control, a means of both bringing the world within their own timescale and putting their experience of the world into a potentially infinite one. Yet these attempts are ultimately unsuccessful in ways that even Men themselves come to recognize, proliferating the death cult and inscribing death rather than long life into their cultural artefacts. In highlighting the Númenórean maps now locked away in the Gondorian archives, Tolkien describes how these maps, despite providing some of the only records available for a now extinct land and civilization, are crumbling into ruin because of 'neglect', as 'all but a few regarded study of what was left of its history as vain, breeding only useless regret for what was lost' (1991, 165). The survivors of the 'Akallabêth' may yearn for their idyllic past, yet the maps no longer provide the comfort of a recognizable and obtainable world and so are ignored. Meanwhile, when exiled Númenóreans believe they can still see the Straight Road to Valinor that was removed from the Earth in the drowning of Númenor, they build very high towers the better to see it, reforming their surroundings the better to see the spatial correlation to their yearning for immortality and unbounded time (1992, 338). Yet as time passes, and the survivors' faith weakens, even these towers are looked upon with scorn. Ultimately, the world cannot be fixed in place and the resultant gap – between subject and representation – only breeds regret and loss.

The central concern of Men is death so that immortality becomes a yearned-for panacea for their troubles and anxieties, yet if Men would look towards the Elves, perhaps they would understand that immortality itself is no promise of a peaceable relationship with time. For the Elves, as much as the Men, have a warped and frustrated experience of their own temporalities, struggling to reconcile their personal experience of time with the world's at large. Tolkien came to rewrite 'The Fall of Númenor' sometime after penning 'The Notion Club Papers'; the resulting text, entitled 'The Drowning of Anadûnê', differs from 'The Fall of Númenor' in crucial ways: here the world is already round and, most importantly in this discussion, the account is what Christopher Tolkien terms an example in Tolkien's experiments in the 'Mannish tradition' (1992, 407), focusing as it does on detailed political and affective experiences of the Númenóreans immediately preceding the drowning of their island. Given this Mannish tradition, the text purposefully betrays, as Christopher Tolkien argues, an 'ignorance of the Elves' (1992, 406); yet contained within it is an exchange which concisely summarizes the tensions in relationships with time that exist both between and for the two races. In the original text of 'The Drowning of Anadûnê', the Avalai (The Undying Ones, that is to say the Elves) speak to the Númenóreans when they sense the growing unrest that will lead to the attack upon Valinor, attempting to quell their fears over the uncertainty of death:

> For we (you say) are unpunished and dwell ever in bliss; and so it is that we do not die, but we cannot escape, and we are bound to this world, never again to leave it, till all is changed. And you (you murmur) are punished, and so it is that ye die, but ye escape and leave the world and are not bound thereto. Which therefore of us should envy the other?
>
> (1992, 346)

Within this exchange is contained what Tolkien explained to Joanna de Bortadano was the 'real theme' of his text: the tension between two experiences of time that both feel like doom. For the Elves, despite their immortality, are not exempt from experiencing the weight of time's passing, nor the anxieties of their own place within time's long movement. In a letter to Naomi Mitchison, Tolkien explains that rather than averting a troubled relationship with non-human temporality, their immortality does not embrace the passing of time but instead freezes it. '[Elves] were "embalmers"', Tolkien writes. 'They wanted to have their cake and eat it: to live in the mortal, historical Middle-earth because they had become fond of it ... and so tried to stop its change and history, stop its growth, keep it as a pleasaunce, even largely a desert, where they could be "artists"' (2006, 197).

There is an eerie similarity here to the Númenóreans themselves, despite their supposedly oppositional experiences of time: the Númenóreans who developed for themselves a tomb culture, learning only how to preserve the dead corpses of their people. The Elves may have immortality, may see themselves naturally stretch into the bounds of non-human time without disruption; yet non-human time is not characterized simply by its infinite scope but also by the change wrought on the physical land. When Playfair grew dizzy with vertigo looking into the 'abyss of time', what he was observing was not a theoretical temporal concept but the very material layers of unconformities in the strata piled up at Siccar Point that indicated billions of years of movement. The Elves may live within non-human timescales, but their yearning for escape, and their refusal to accept the changing of the Earth within the slow, extended scale of deep time, disrupts non-human time as much as Men's longing for immortality, problematically recentring anthropocentric concerns in a timescale that should be independent of them.

We see this in the enchanted realm of Lothlórien, the home of Galadriel and the Galadhrim, which is caught up in a liminal temporality between the mortal world of Men and the immortal world of the Elves; both a physical space that can be entered and negotiated and which borders onto other, mortal spaces, and also a space that unsettles the passage of time that the rest of Middle-earth undergoes. 'The love of the Elves for their land and their works is deeper than the deeps of the Sea' (2008a, 475), Galadriel tells Frodo and Sam when they look into her mirror, and the traces of this love are felt throughout Lothlórien, a land frozen within an idyllic, untouched past. The experience of crossing into Lothlórien reads similarly to crossings in portal fantasies: there is the sense of a threshold being traversed, and a new world discovered. Frodo senses this shift when he steps onto the banks of the Silverlode: 'it seemed to him that he had stepped over a bridge of time into a corner of the Elder Days […] in Lórien the ancient things still lived on in the waking world' (2008a, 454–55). The disruption of the linearity of time is quite evident; Lothlórien does not recall the past, or even actively recreate it, but rather exists *within* it so that the passage of time has effectively been halted within its boundaries. Elsewhere, Haldir says Cerin Amroth – a mound in the heart of Lothlórien – 'is the heart of the ancient realm as it was long ago' (2008a, 456), and Frodo feels that 'he was in a timeless land that did not fade or change or fall into forgetfulness' (2008a, 457). The past is pulled into the present as if it never left, the supposedly ancient made immediately tangible and contemporary. The disruption of linear time is also seen in the collapsing together of different tenses: Aragorn re-experiences his first meeting with Arwen, to the extent that he momentarily sees Lothlórien as it was on that day and his own appearance briefly changes to that of his younger self, while Galadriel's mirror simultaneously shows the beholder 'things that were, and things that are, and things that yet may be' (2008a, 471).

This temporal dislocation is in many ways a classically sylvan characteristic: because forests lie beyond ideas of linear time, Robert Pogue Harrison argues, 'a protagonist wandering through a forest experiences a terrifying or enchanting loss of temporal boundaries, as if he or she has passed into a world of implications which render our deepest structural categories superfluous or unreal ...' (1992, 8), so that Lothlórien's resistance to temporal linearity in many ways builds on cultural conceptualizations of forests as spaces void of human rational structures of time and space. Yet there is a specific cultural aspect to the Elves' unnatural preservation of Lothlórien that speaks to relationships with time that lie beyond typical depictions of magical forests, for what is this forest frozen in time than an act of embalming, a 'corner of the Elder Days' (2008a, 454) that ought to be relegated to the ancient depths of deep time but is not. Lothlórien's timeless quality is not the non-human world resisting human rationalizations of time, but a spatial representation of the Elves' anxieties around and resistance to time's natural passing, an imposed denial of the tension between their lifespans and the Earth's, and a culmination of their both wanting and eating their cake. The conversation between Frodo, Sam, Aragorn and Legolas after they leave Lothlórien is indicative of this intrusion. Sam comments that it was as if time did not 'count in there', the word 'count' having, as Flieger remarks, the ambiguous meaning of both 'to matter' and 'to add up' (Flieger 2001, 93). Legolas then confirms this, his explanation leaning towards the latter meaning: '[Elves] do not count the running years, not for themselves' (J. R. R. Tolkien 2008a, 506). Yet the former meaning is also implicit in his words; the Lothlórien Elves pretend that the passing of the years do not matter, at least not for them, and they pull the physical realm of Lothlórien into this deceit.

Of course, there is a certain protective element to this not counting. Lothlórien acts as a safe haven for its inhabitants from the evils of the world, and its nature is secured from decay and harm; the preservation of the mallorn trees, for example, stands in contrast to the destruction of other trees throughout Tolkien's legendarium. Earlier drafts of *The Lord of the Rings* make the extent to which Lothlórien can be read as a safe haven even more explicit: the narrative describes how '[e]vil had been heard of ... but it had not yet stained or dimmed the air', and that despite the winter, 'nothing was dead, only in a phase of beauty ... there was no smell of decay' (1989, 241). And certainly, the Galadhrim and the Fellowship find beauty and respite in their surroundings: Aragorn's memories here are 'fair' (2008a, 458) – no wonder, when there is so much splendour to be found – and Frodo feels his breath taken away at the first sight of Cerin Amroth. The woods of Lothlórien are neither evil nor dystopian and yet, as Tolkien stresses, the creation of such timeless protection comes at a great price, constructing a world which is 'frozen' and 'embalmed'; that is to say, a world which cannot shift and wear and evolve according the natural life cycles of the Earth. For Tolkien, the preservation of the forest is crucially less rooted in environmental concerns than it is in the Elves' desire to preserve their place in and experience of the world: they want the 'peace and bliss and perfect memory of "The West"' that they experienced in Valinor (2006, 151), yet recreated on Middle-earth where they are considered the superior beings. The enforced maintenance of Lothlórien becomes another manifestation of the control over the non-human world – its time and space – that is practised by the humanoid creatures of Middle-earth; its unnatural preservation exploits the natural rhythms of the non-human world to give the Elves pleasure and a sense of control over their immortality, bringing their geographical spaces into deep time while denying all the effects that deep time ought to have. The resulting uncanniness underlines the unsettling nature of such disruption: 'Alas it is winter' (2008a, 439) sighs Legolas, when first they set foot in Lothlórien, but on their first waking morning in the forest the 'sun of a cool summer's morning' spreads over the land (2008a, 450). The Fellowship experience the 'cool ... soft' air of early spring at the same time as the profound 'quiet of winter' (2008a, 466). It is hard for Sam to tell if 'they've made the land, or the land's made them' (2008a, 469), signifying a particular collapse of boundaries that embodies the control exercised over Lothlórien's temporality. Tolkien stresses the full effects of this unnatural state in a conversation Treebeard has in *The Two Towers*, when he explains to the hobbits that Lothlórien is 'fading, not growing' (2008e, 608), and by Galadriel, who acknowledges that eventually the 'tides of Time' will sweep Lothlórien as it currently stands away (2008a, 475). Nothing, beautiful or evil, can emanate and grow from this artificial stagnation, Tolkien stresses. Ultimately such embalming – much like the preservation of the dead Númenóreans – is but a crystallization of the moment of death and acts only as a reminder of the Elves' frustrated experiences with their immortality: their prolonged lives in a world to which they are bound, an excruciatingly slow anticipation of their own fading.

How does one map a land like Lothlórien, a land so entirely bound up in the anxieties of its inhabitants that its very nature has changed? 'Without a doubt it is meant to be a real place. It is on a map,' Flieger explains (2001, 91). Yet for all Lothlórien is a 'real' place, it no longer exists within 'real', natural temporalities, and maps (as we have seen) are as much temporal signifiers as they are spatial, if not of accuracy then of experienced reality. Yet perhaps the cartographic representation of Lothlórien is the most temporally accurate of all of Middle-earth's maps, for in the artificial fixity of the map as object is found the same stagnation that characterizes Lothlórien's landscape. The map becomes a tangible manifestation of the enchantment and control cast over Lothlórien; nestled on the Middle-earth map between the Drimrill Dale and the Field of Celebrant, the Nimrodel running neatly through it in ink, Lórien is preserved forever, a cluster of trees that – much like the mallorn themselves – can never wither or fade. The tension that Tolkien crafts between synchronicity and asynchronicity in Lothlórien mimics the same tension between the two contained in the map: the pretence towards temporal thickness and the map's snapshot-like quality, the flattening of deep time onto a single sheet of paper and the prolonging of a single human experience of the non-human through an unchangeable artefact. Map-making, and in particular constructing the temporalities of map-making, is an inherently paradoxical act, and nowhere is this clearer than in the small inked-in forest of Lórien, an expression not of the landscape but of the anxieties of those living within it.

As usual, Tolkien's maps tell us more about the map-maker than they ever could about the landscape they supposedly depict. True, the preserved-in-amber quality of Lothlórien mimics that of its depiction on the map, making it a decidedly 'cartographic' land. Yet the map fails to convey the discordance in time between it and the rest of Middle-earth, instead representing both – through its own temporal flatness – as equally timeless. The whole of Middle-earth's cartography, much like the forest of Lothlórien, exists in past, present and future all at once: an expression, perhaps, of the longing of the Elves to embalm their experience of the world as their time stretches on forever, of the yearning of the Men for the immortality that this temporal collapse would imply. Ultimately, it is the inability of both races to reconcile themselves to the natural movement of non-human time and their existence – whether prolonged or not – in deep time that is the cause of such profound fear and regret. 'Which therefore of us should envy the other?' the Avalai ask (1992, 346). Yet there is no peace to be found, Tolkien shows, in either position.

Mapping Anthropological Change

Both Men and Elves resist the passing of time, and both envy the other's position, yet ultimately the tragedy of both races is that they must eventually leave Middle-earth, whether individually through death or as a species through fading. As with the Númenor maps and their depiction of a now destroyed land acting as a window into the past, the maps of Middle-earth can be vehicles for historical and specifically anthropological contemplation. They are a record of history, except they preserve not only the geology of the past and experiences of the physical world but also the traces of species who peopled its lands.

The analogy between maps and fossils is again pertinent here, especially in regards to the Elves, whose experience of leaving Middle-earth acts effectively as an extinction. As Rateliff argues (2006, 67), the framing of Middle-earth as a prehistory of our own world implies the death and extinction of all the species that no longer exist in it, and in the case of the Elves, this extinction is made explicit throughout the texts, both through the act of 'fading' and through their departure from Middle-earth to the Undying Lands. The former is a process that all Elves naturally undergo: although originally the Elves were intended to have both immortal bodies and spirits, the evil that enters Aman thanks to Morgoth causes an eventual consuming of their bodies, or hröa, by their spirits, or fëa. Thus, although Elvish bodies are capable of withstanding disease, injury and ageing and last for several Ages of the world, eventually the dominance of their fëa increases, so that '[a]s the weight of the years, with all their changes of desire and thought, gathers upon the spirit of the Eldar, so do the impulses and moods of their bodies change' (J. R. R. Tolkien 1993, 212), leading to the spirit 'consuming' or fading the body. The immortal fëa then enters the Hall of Mandos in the Undying Lands, where it waits to someday be reborn. This waning or fading can only be entirely avoided by leaving Middle-earth and returning to Valinor, where the Elves could remain both immortal and incarnate. Elves inhabiting Middle-earth are imbued with sea-longing, seen in Legolas when he spies gulls flying above Minas Tirith and is suddenly filled with an unquenchable desire to cross the sea: 'their wailing voices spoke to me of the Sea. The Sea! Alas! I have not yet beheld it. But deep in the hearts of all my kindred lies the sea-longing, which it is perilous to stir … No peace shall I have again under beech or under elm' (2008c, 1143). Legolas' sudden drive to leave Middle-earth speaks to Galadriel's prediction that 'we must depart into the West, or dwindle to a rustic folk of dell and cave, slowly to forget or be forgotten …' (2008a, 475).

The slow death of their bodies is inextricably connected to movement through the physical lands of Middle-earth and beyond: Legolas' instinct towards the sea is not merely his own longing, but part of an ingrained mass movement of his people to preserve their lives by passing out of Middle-earth. His emotive apostrophe to the sea sits strangely alongside his immediate acknowledgement that he has not even seen it; its significance is a biological imperative rather than an affective attachment.

Whether through the slow overwhelming of their bodies by their fëa or their mass exodus from mortal lands to the lands of Valinor beyond the sea, the Elves will slowly lose their place in Middle-earth, no matter how much they preserve their woods in time. Whichever way it transpires, Galadriel recognizes that the power of the Elves is weakening, and their time in Middle-earth is drawing to a close, a waning of power that is explicitly linked to the ensuing dominion of Men in the land: the slow extinction of one peoples, the increasing stronghold of another. Tolkien first refers to anthropological shift in his early writings featured in *The Book of Lost Tales I*. In a short introduction to Tol Eressëa he writes, 'so it is that the Magic Sun is dead and the Lonely Isle drawn back unto the confines of the Great Lands, and the fairies are scattered through all the wide unfriendly pathways of the world; and now Men dwell even on this faded island, and care nought or know nought of its ancient days' (1983, 25). At this point in his mythology, Tolkien does not portray the fading of the Elves as an extinction from the world, but rather stresses the loss of their land and culture and the concomitant loss of their sense of belonging: much as their home becomes physically subsumed into the Great Lands that are dominated by Men, so too is their culture neglected and overwritten by the Men that come after and who neither know nor value the knowledge of this Elvish past. Christopher's commentary on this section makes explicit the causal effect the coming of Men has on the Elves' diminishment: 'Men entered the isle, and the fading of the Elves began', he explains, framing it not only as a scattering through a changed world but as a specific decline (1983, 26). This particular act of fading is then expanded on throughout later writing in the legendarium: in the 'Quenta Silmarillion' draft in *The Shaping of Middle-earth*, Tolkien describes how Lúthien faded in much the same way the Elves of later days faded, 'when Men waxed strong and usurped the goodness of the earth …' (1986, 134), while a footnote referring to Men in the 'Quenta Silmarillion' draft in *The Lost Road* reads: 'The Eldar […] named them the Usurpers, the Strangers …' (1987, 245). The replacement of the Elves by Men in the hierarchy of Middle-earth is a lamentation that weaves throughout the legendarium, stressing the fading of the Elves from the mortal lands not merely as a natural fact, but as a domination of one group over another. By the time of the events in *The Lord of the Rings*, meanwhile, this fading is almost complete. Permeating the quest story of the hobbits and the ring is a pervading sense of loss, a melancholy elegy for a dying age. Gandalf speaks to Aragorn about how 'the time comes of the Dominion of Men, and the Elder Kindred shall fade or depart' (2008c, 1272); the Appendices discuss how the Third Age was synonymous with the 'fading years of the Eldar' (2008c, 1272); and the Prologue makes clear that by the beginning of the Fourth Age, the last of the High Elves – Elrond, Celeborn and Galadriel – had departed Middle-earth, leaving it to the race of Men.

The emphasis on the slow decline of one race and the increasing dominion of another frames the fading of the Elves not merely as an existential experience, but an anthropological one. Much as the physical land of Middle-earth wears and alters through the passing of the ages so too does the movement of its people, a slow anthropological shift that occurs over Middle-earth's long abyss of time. Yet as with any slow change across time, the alteration of Middle-earth's population and the slow extinction of the Elves from the mortal lands has a necessarily spatial aspect. As Jason Fisher observes, ideas of mortality and immortality in the legendarium are frequently conveyed through spatial metaphors, once more binding conceptualizations of time – both bound and unbound – to movements through space (2010, 3). Tolkien's 'Circles of the World' trope, Fisher argues, is a key example of this: ostensibly referring to the physical land of Arda, the phrase comes to signify both the material limits in which Men and Elves dwell while in Arda, as well as the promise of something more beyond, a gesture towards a kind of liberated immortality that exists beyond the regulated space of mortality. Speaking to his wife Arwen on his deathbed, Aragorn invokes the phrase to comfort her sorrow at his passing. As Vaninskaya argues, their exchange represents a reversal in the typical positions of Men and Elves regarding mortality, with Arwen finally understanding the validity of Men's fear, and Aragorn the beauty of faith in the unknown (2019, 171). As Arwen laments at the bitter gift of death that her husband has been given, Aragorn urges her to sorrow rather than despair. 'I speak no comfort to you, for there is no comfort for such pain within the circles of the world,' he tells her, yet adds, 'Behold! We are not bound for ever to the circles of the world, and beyond them is more than memory' (2008c, 1394). Despite Aragorn's claim that he can speak no comfort, there is indeed comfort in his farewell, an invocation of the existential possibilities that lie beyond the physical and metaphysical properties of Arda. For Fisher, the recurrent imagery of the Circles of the World is a physical manifestation of the nostalgia and loss that threads its way throughout the legendarium, the continual acts of rupture between the physical world and its inhabitants and the simultaneous necessity of coming to terms with this rupture (2010, 2). The fading of the Elves spans this doubled spatial and metaphysical dimension: when the Elves leave the 'Circles of the World', it is both a reference to a physical act and to their transcendence of the world's material and mortal limits.

Yet as much as Tolkien's evocation of the 'Circles of the World' speaks to both a material boundedness in the mortal lands of Arda and a potential material unboundedness that lies beyond, there is nevertheless a markedly spatial, and specifically geographical, aspect to the Elves' slow extinction and diminishment to the immortal realm of the Undying Lands. Tolkien casts these lands in emphatically geological language, emphasized particularly in the 'Ambarkanta', which depicts Valinor's mountains, shores and sloping lands with topographical specificity: it is described as not far from the Walls of the World, with mountains that 'curve backward' and a Western shore that is at the level of the sea (1986, 239). This physical characterization is stressed again in 'The Fall of Númenor', in which Valinor is 'sundered from the earth' (1987, 15), once more casting the separation of the mortal and immortal lands – the Circles of the World and beyond – in tangible, material terms. And even after the removal of the Undying Lands from the Circles of the World, a spatial manifestation of the connection between the two lingers in The Straight Path, a path of access that can only be crossed by Gods and Men that exists between the metaphorical and physical. Its description in 'The Fall of Númenor' is very hesitant: the old line of the world is described as a 'memory' that endures, and the path that remains is 'likened' to a plain of air, a straight vision and a bridge (1987, 17), without ever establishing its nature in more concrete terms. Yet, much like the Undying Lands it connects to, the path also has a certain physicality: boats sail through it from the Grey Havens, and it is said that Elves and Gods can walk on it.

The slow extinction of the Elves, their fading trajectory through the long Ages of Middle-earth and their departure from the Circles of the World over the ages therefore have a physical and geographical manifestation in this road and the Undying Lands. Both Valinor and the road are represented in those of the maps which depict Arda as a whole, their physical presence made all the more tangible through their cartographic representation. The first is the I Vene Kemen map, found in *The Book of Lost Tales*, where Arda is depicted in the shape of a ship and Valinor is labelled near one of the helms. Maps IV and V of the 'Ambarkanta', meanwhile, feature Valinor as a land mass towards the extreme west, with recognizable topographical features such as the mountains shielding it from the rest of the world. Diagram III of the 'Ambarkanta' maps features a line at a tangent to the circular world, cutting through the layers of water and air arranged in concentric circles around. This line is labelled 'The Straight Path', and as this map is titled 'after the Cataclysm', it is almost certainly the path to the Undying Lands. The early maps from the First and Second Age are straightforward depictions of the Earth, at a time when the Undying Lands were an undeniably physical place that could be mapped much like anywhere else (and accessed, as the assault from the Númenóreans shows all too well). Yet Diagram III, explicitly made after the cataclysm, suggests a physicality that is not immediately obvious from Valinor's physical and metaphysical separation from the corporeal and finite lands of Arda. The appearance of 'The Straight Path' on the map acts effectively as a cartographic representation of the physical and spiritual journey of the Elves from the mortal 'Circles of the World' to the culmination of their fading and their return to the immortal lands of Valinor: a spatial representation of the temporal process of their extinction from Middle-earth.

Valinor is removed from the Earth, yet its physical presence lingers; it is no longer part of the Circles of the World, yet the path that might take you there (but never back) is depicted on the map. The lands of Middle-earth are haunted by past assemblages of landscapes that came before, landscapes that speak to past ways of life and possible future ways that were ached for and always beyond reach. Born from the Age of Anxiety that he himself was writing out of, Tolkien's characters experience the same destabilization of their temporal and existential position within the vast and indifferent mechanisms of the Earth. When the Númenóreans murmur against the decree of their relatively short lives, they are really agitating against the impeding finality of human time in the face of non-human time: a temporal scale for which they long but, as Michael Ramer of 'The Notion Club Papers' knows all too well, would only be unendurable. Ultimately, Tolkien understood that accepting one's place in the world means accepting death: not through the creation of a death cult but, like Aragorn, through an act of faith that something else might be waiting, and a letting go. Without this, it is impossible for humans to accept their place in the world, to accept the bounds of human time and the inaccessibility of non-human time. The world is rent apart, temporalities collapse, and death and annihilation of both the human and non-human take place.

The haunting of past landscapes torn apart out of greed and fear is, in a way, what Tolkien's entire legendarium philosophically resists, and what maps attempt to entrench and legitimize. Through his crafting of aeonic timescales and non-human temporalities, his construction of fragile yet resisting geologies that are all too vulnerable to human anxieties, Tolkien's keen understanding of this kind of haunting becomes entirely evident. He was, as Flieger has argued, a man of his time, writing through a crisis of modernity. Haunted by his own present and the effects of the industrial boom, he was also, perhaps, haunted by his futures: by the things he suspected and feared would come to pass. Looking back now, over decades of Anthropocenic destruction and deep-time collapse, it is startling how much of it has come true.

4

Imperialism and Race

Throughout this book, I have argued that Tolkien embeds a radical approach in his literary engagement with the politics of land; that he was a man, in many ways, ahead of his time. This is true, and also not true, of his approach to empire and particularly to race – one of the most significant exercises of power over land that scaffolded his social and political context. Tolkien's engagement with race is an ongoing complicating factor in his legacy as an author, to the extent that the discourse around the subject has collapsed into simple, and unproductive, binaries of ascertaining whether his writing, and by extension he himself, were racist. This has led to a certain level of defensiveness in the scholarship around this subject, which works both to erase the explicitly wrong ways in which he approached ideas of race and colonization, as well as the more complex interventions he made into a surrounding imperialist context with which he was markedly uncomfortable.

Yet to fully understand Tolkien's engagement with land and environment requires a clear and unabashed confrontation with his engagement with race and empire, for narratives of environmental and imperial harm tell entirely the same story. That the effects of empire are an ecological as well as racial concern is something that has become increasingly clear as the climate crisis continues to rage: both its roots and aftershocks are marked in the clear history of racial exploitation that marks out the history of the world. Of course, the Anthropocene is a fundamentally ecological and geological affair and it is essential that it is focalized as such: it is only in liberating the non-human – flora, fauna and mineral – from the toxic hierarchies of human cultural thought and recentring it within the concerns of the Earth that we have any hope of escaping mass environmental destruction, a state of matters that Tolkien understood only too well. Yet the legacies of exploitation, extraction and neglect that have come to define the histories of our lands, both fictional and real, are inextricable from the experiences of the humans who live on them. Exercises of power over land inevitably harm its inhabitants, and indeed drive home just how mutually entangled and interdependent the two are – however much we would sometimes like to pretend otherwise.

As Kathryn Yusoff argues, the critical field of Anthropocene studies so easily lends itself to depoliticization (2018, 2), framing the origins of this age of mass ecological and geological upheaval as an inevitability – negative, to be sure – of human hubris and aspiration towards technology and mastery in the years of the Industrial Revolution, modernization and capitalist entrenchment. Yet the catalyst for the Anthropocene, as Yusoff determines, is inextricable from the histories of imperialism and colonialism that mark out our collective pasts and that fundamentally enabled the industrialization and modernization of the West and our models of capitalist wealth accumulation. This is, to be clear, no mere abstracted or metaphorical claim. The genocide of Indigenous populations following the invasion of the Americas by European settlers, for example, led to a mass population decline from 54 million to 6 million over a century and a half, a staggering loss that was also observable in the decline of carbon dioxide levels in the Antarctic ice cores due to abrupt reductions in sustainable farming and the subsequent regrowth of forests (2018, 31). People – particularly marginalized people – and the Anthropocene are bound up in ways that inscribe histories of racism, violence and exploitation of humans by humans into the Earth's geological and ecological record. This is not intended as a recentring of the human within Anthropocenic discourse, but an acknowledgement of the ways in which certain peoples – Black, Indigenous, enslaved, racialized – have long been absorbed into the category of the non-human, or more explicitly, the inhuman. I wrote in the second chapter of this book about the centuries-unchanged hierarchical binary of human/non-human that laid the groundwork for the Anthropocene to emerge; it is essential, I believe, to understand the same violent extraction that the not-considered-human undergoes within these processes.

This mutual vulnerability of the land and its inhabitants, often (although not always) within and produced by a severely racialized framework of relations and power, underwrites issues of environment and land in Tolkien's legendarium in a similarly prescient way to his other examinations of land politics, and is explored with varying levels of both astuteness and ignorance. The principle narrative tension across his literary output ostensibly distils down to the battle between good and evil – the Ainur, the Elves and Men alternately battle Morgoth, Sauron and Saruman – yet this continued moral struggle largely manifests as a struggle for land: who physically and psychically occupies and controls lands of Middle-earth, Beleriand and Valinor. This is made evident both narratively and lexically: in *The Lord of the Rings*, Gandalf warns Frodo that with the One Ring, Sauron will finally have the power to 'cover all the lands in a second darkness' (J. R. R. Tolkien 2008a, 67); in *The Hobbit*, the quest to defeat the evil dragon is implicitly connected with the reclaiming of the Lonely Mountain and its surrounding lands; and in the 'Silmarillion' writings, the degree of success or defeat in the confrontations between Melkor and the Valar is determined by the shifting occupation of territory. Throughout Tolkien's legendarium, the landscape is a contested object; it is both pawn and prize in the conflict between Middle-earth's forces, and the struggle for power frequently equates to a struggle for authority over and ownership of land.

This chapter intends to tease out the complex colonial, imperial and racial underpinnings of Tolkien's legendarium and political thought: the ways in which he clearly understood colonialism and empire as an exercise of power over land, the ways in which he understood that such structures are inextricable from the broader ecological concerns of our meddling in and on the Earth. Looking at representations of defence, conquest, empire and colonization in the legendarium, this chapter examines the ways in which the act of cartography reflects and empowers this politicization, and the imbrication of environmental exploitation within the maintained vulnerability of a land's inhabitants. Beginning with a short overview of postcolonial discourse as a means of conceptualizing human and non-human proximity to harm, it considers the politicization of land in Middle-earth, the significance of threshold and border spaces in demarcating both the boundary of the land and the boundary of power over land, and acts of small- and large-scale defence and conquest that draw out human and non-human precarity and the narratives of racialization that work to complicate certain of these acts.

Rather than commenting on Tolkien's world-building strategies or drawing out historical allegories in his territorial relationships and conflicts, I want to position Tolkien's depiction of land politics as a demonstration of the ways in which he was attuned to the various mechanisms of violence that participate in anthropocentric and Anthropocenic ecological and geological destruction. Although not allegorically the closest model of empire to Middle-earth's own,[1] reading Tolkien alongside postcolonial theory responding to modern Western empires reveals the ways in which he was participating, both successfully and

---

[1] See James Obertino on the specific structural and episodic affinities between Tolkien's legendarium and ancient and medieval imperial histories.

unsuccessfully, in a demonstrably modern critique of the entanglement between human and non-human degradation and racialized violence that is inherent to the strategies of modern Western empires, placing him in conversation with contemporary concerns surrounding imperialism, British Empire, and the intimate harm of land politics. Borrowing Edward Saïd's definition of imperialism as 'the practice, the theory, and the attitudes of a dominating metropolitan center ruling a distant territory' and of colonialism as 'the implanting of settlements on distant territory ...' (1994, 8), I intend to examine how power over land in Tolkien's legendarium operates both within these fixed structures of oppression and extraction as well as within unstructured skirmishes and invasions that reveal the inevitable damage that exertions of power over land can have, both ecologically and socially. Reading Tolkien, I am struck continuously by the contradictions of his approach: the moments where he edges towards a strikingly anti-colonial mindset, the moments where he falls back on the harmful racialization that characterized, and indeed fuelled, the gutting force of the British Empire. To place Tolkien within a context is to place him inevitably within an imperial context, which is to say, inevitably within a racist context. What can looking fearlessly and unbiasedly at his work tell us about the necessities of understanding the entanglement between human and non-human harm? What can it tell us about an author who was suspended, constantly, between past, present and future?

The Politics of Land and Map

Postcolonialism is, by necessity, a critical field that is preoccupied by questions of land, concerned as it is predominantly with 'displacement' (Nixon 2005, 235) and thus the absence and denied possibility of emplacement. In his examination of the psychological, social and cultural trauma of colonialism in *The Wretched of the Earth* (1961), Frantz Fanon points to land as the crux of colonial conflict for both sides: colonizers desire expanded power over lands that do not belong to them, while for the colonized, 'the most essential value, because *the most concrete*, is first and foremost the land: the land which will bring them bread and, above all, dignity' (1963, 44, emphasis added). Although he is speaking predominantly to the social and individual rather than ecological effects of colonialism, Fanon nevertheless positions land – not the abstract conceptualization of land as nation, but the material reality of land as habitation, resource and psychic comfort – as central to the violent legacies of settler colonialism and dispossession. For Édouard Glissant, this tension between land as site of habitation and connection and land as political and economic potential is at the heart of the colonial project. Colonialism, neo-colonialism and invasion as means of land expansion reform cultural and psychic relationships with the land, transforming it into territory that fundamentally embeds conflict within the relational possibilities of the physical landscape. Territory, Glissant argues, is 'the basis for conquest ... Territory is defined by its limits, and they must be expanded. A land henceforth has no limits' (1997, 151). Colonial imaginaries render land into nation or territory or conquest, embedding the violence of dispossession, extraction and oppression within its physical space. As Saïd states so plainly, all of human history is 'rooted in the earth' (1994, 5), reifying the political signification of land through a material, terrestrial image that draws stark attention to the significance of its physical presence within abstracted narratives and acts of power. 'The main battle in imperialism is over land, of course' (1994, xiii), explains Saïd: the urgency, intentionality and activity of settler–colonial imperialism cannot occur without 'thinking about, settling on, controlling land that you do not possess' (1994, 5).

This stress on the physicality of land operates at two levels. First, it makes explicit its centrality not only to the practicalities of the imperialist and colonialist projects but to human experience; as Saïd says, 'none of us is completely free from the struggle over geography' (1994, 6), a struggle that Saïd notes manifests just as much in our cultural markers of belonging, identity and imaginations of selfhood as it does in soldiers and warfare. Second, it draws attention to the doubled dispossession of both a land and its people; the inextricable simultaneity of human and non-human extraction in colonial and imperialist conquests 'over land and the *land's people*' (1994, 6, emphasis added). The land and its native inhabitants are intimately connected through this experience of occupation and trauma; imperialism as an expression of power, exerted through cultural hegemony and prescribed narratives, begins in the conquest of physical space. Imperialism thus becomes an 'act of geographical violence' – a violence committed both *to* and *through* geography – so that the colonized subject's experience of subjugation is initiated through the loss of the physical signification of their psychological sense of belonging (1990, 77).

Although Fanon, Saïd and Glissant's critique is rooted in the social, cultural and political significance of the land as physical entity, their investigation into empire and colonization nevertheless eschews a specifically ecological grounding, largely framing the non-human world and landscape through the lens of human habitation, identification and ownership. This epistemic gap is typical of what

Rob Nixon terms the 'broad silence' (2005, 233) that exists between the fields of postcolonialism and ecocriticism: two critical disciplines that engage both simultaneously and historically separately with the political, ontological and psychic signification of land. For Nixon, this gulf is caused and exacerbated by the superficially oppositional approaches the two disciplines take: postcolonial writing being occupied as it is by ideas of 'displacement' and ecocriticism by 'literature[s] of place' (2005, 235). Yet disrupting the rigid boundaries held up between the fields allows for a 'more historically answerable and geographically expansive' constitution of our environment and environmental thought (2005, 247), allowing us to recognize not just the concurrent exploitation of the non-human and its human inhabitants within imperialist processes, but the deliberately conceived and maintained structures of power that designate both the non-human and particular humans – Indigenous, racialized – as inhuman. The non-human is recognized as part of the subaltern subject of the imperialist strategy, one that is – as Elizabeth DeLoughrey and George B. Handley explain – 'a participant in this historical process rather than a bystander to human experience' (2011, 4). The environmental cost of imperialism, seen in acts of extraction, profiteering and heedless expansion that result in pollution, desertification and deforestation, becomes a casualty of imperialism in its own right, rather than merely representative of or heralding the dispossession of native homeland. Land thus becomes 'crucial as recuperative sites of postcolonial historiography' (2011, 8): evidencing histories of exploitation long after the imperialist project has repressed and silenced its acts of violence in manipulated human narratives.

This slippage between the colonized human and non-human goes both ways. The non-human is made subaltern, but so too is the human rendered inhuman. As Graham Huggan and Helen Tiffin argue, imperialism and colonialism are fundamentally based in a mutually conducive environmental racism, so that the oppression of the non-human is both inextricable from and strictly enabled by the racial oppression of the human colonized subject. Native inhabitants are racialized as primitive, uncivilized and animalistic – much like their environment – and assimilated into an ideology that justifies and validates their exploitation and legal and extralegal genocide. In this way, the history of the Anthropocene, or environmental harm more broadly, becomes inextricable from histories of racial violence; indeed, the history of environmental harm *is* a history of racial violence. This porous proximity between the two is central to Yusoff's imagination of the Anthropocene, as already briefly touched upon, which demands a geological understanding of race and a racial understanding of geology in order to open up the entanglement between relations of human and non-human extraction. Crucially, Yusoff underscores how each of the proposed Golden Spikes, or start points, of the Anthropocene epoch can be traced back to the intimacies of harm absorbed by Black and brown people during moments of notable geological and ecological change: the genocide of Indigenous populations in the colonization of the Americas in 1452; the critical role of slavery in fuelling the Industrial Revolution (so often masked in Anthropocene discourse by the somewhat twee image of the invention of the steam engine); the displacement and death of Pacific populations during the nuclear testing of the Manhattan Project. As Black Studies scholar Christina Sharpe argues, there is a very clear wake – both racial and ecological – left behind every history of trauma and dispossession. Sharpe's examination is striking in its own explicitly ecological language: 'In what I am calling the weather', she explains, 'antiblackness is pervasive as climate. The weather necessitates changeability and improvisation; it is the atmospheric condition of time and place; it produces new ecologies' (2016, 106). Of course, Sharpe's utilization of ecologies and weather is to an extent symbolic, but it also provokes a vital question on the relationship between human and non-human vulnerability in the face of imperialist histories. If frameworks of sustained violence become the weather, how does this come to affect our ecologies, both metaphorical and literal?

As a method of representing and narrating the land, cartography is also complicit in contributing to and perpetuating these hierarchies of power. As argued throughout this book, mapping is an inherently political act that is fundamentally concerned with the maintaining and exerting of power over both the land and the land's inhabitants, yet the explicit and implicit biases and perspectives of the map are often disguised by its pretence at objectivity. As Wood and John Fels argue,

> [t]he dominant view of modern Western cartography since the Renaissance has been that of a technological discipline set on a progressive trajectory. Claiming to produce a correct relational model of terrain, maps are seen as the epitome of representational modernism, rooted in the project of the Enlightenment, and offering to banish subjectivity from the image.
>
> (2008, 6)

With the advent of scientific surveying techniques and new representational technologies, the post-Enlightenment map became an aspirational model of objectivity and impartiality. However, in practice, the map cannot be separated from the political environment that produced and uses it. Wood highlights that all maps 'inevitably, unavoidably, necessarily embody their author's prejudices, biases, and partialities' (1992, 24), while Harley describes maps as a 'way of conceiving, articulating, and structuring the human world which is biased towards, promoted by, and exerts influence upon particular sets of social relations' (1988, 278). Both Wood and Harley position the map as a form of text that imbibes, encodes and projects a particular politics: seen in what the cartographer chooses to represent and what they choose to omit, in the map's reinforcement of potentially contested spaces such as borders and territories, and in its use of paratextual legends to guide the reader through the map. Moreover, it is not only the system of production that politicizes the map, but also how the map is deployed. From attesting property rights over private pieces of land to exploring and claiming new lands during imperial expansion and warfare, maps are used to demonstrate ownership of and power over land.

Imperialist mapping in particular highlights the extent to which cartography can be used as a political tool, and the intersection between the conquest of land and the conquest of people. Edney examines how maps were used in the nineteenth-century British imperialist conquest of India, arguing that the mapping of Indian territory became an extension of the geographical violence of imperialism, where every inch of the land is examined, calculated, valued and brought under a new political – and in this case textual – control (1997, 24). Edney argues that imperial mapping works to recreate and reaffirm the empire in another medium, 'subsuming all individuals and places within the map's totalizing image …' (1997, 24). The imperialist project therefore occurs twice: first, in the claiming of power over the lands, and second in the claiming of a complete and encompassing Foucauldian knowledge of the lands through their representation. Harley reinforces this argument, emphasizing that imperial mapping was not only a practical tool for gaining knowledge and control over unknown spaces, it was also 'used to legitimize the reality of conquest and empire' (1988, 282). Maps become a form of hegemonic and cultural domination: in every instance of mapping, there is a knowledge of the territory, and in particular an assumed truth about the territory, which the map supposedly conveys. The map and its makers are placed in the privileged position of the historian in Foucault's historiographical critique: they are gatekeepers to the land and its cultural representation, a material evocation of imperialism's strategy of physical, cultural and psychic dispossession.

(Dis)possessing Middle-earth's Lands

Before turning to the distinctly ecological and racialized forms of dispossession that Tolkien draws out in Middle-earth's various land conflicts, it is worth examining more broadly the ways in which land in Middle-earth is, as Glissant lays out, territorialized and opened up to narratives of human and non-human exploitation. For a land to become dispossessed it must first, of course, be possessed, a state of affairs only too familiar in Middle-earth's severely administrated, patrolled and demarcated land mass. Whether through formal governance or a personal sense of affinity with – and thus claiming belonging over – the land, there are various models of territorial possession which work to politicize the lands of Middle-earth, transforming them into an object with both affective and exploitable value. These models, as discussed in the second chapter's context of human/non-human encounters, lie on a spectrum of unfettered exploitation and instrumentalization to gentle mutual dependence and attachment, yet – as Tolkien himself states – each remains predicated on an indisputable measure of control drawn from and over the land (2006, 178–79). A lot of these proprietary, hierarchical relationships have been discussed at length in the previous chapters; yet it is worth briefly touching on two examples here that epitomize the transformation of, as Glissant puts it, land into territory, and the potential this opens up for politicization and exploitation of the human and non-human through land conflict.

The first, and most recognizable, evidence of such politicization lies in the large, feudally constituted kingdoms of Middle-earth in the Third Age: Gondor, Arnor and Rohan. These areas mimic medieval administration of power, where different regions such as Dor-En-Ernil and Ithilien are ruled by princes who are given power over the land by the king and who remain answerable to him. The land is thus populated, controlled and possessed through multiple levels of authority, lending every aspect of land relation – habitation, agriculture and extraction, and governance – a distinctly political function. The establishment of the kingdom of Rohan concretizes this use of land as a means of building and maintaining power relations. Its origins lie in a complex network of conquest and allyship: when a host of wild men from the North-east and orcs from the Misty Mountains converge on Calenardhon (a region of Gondor) in the Third Age, the Gondorians call for aid from their allies the Éothéod, a race of Northmen led by Eorl. After the Éothéod help the Gondorians defend their lands against the invaders, Gondor's Stewart Cirion gifts Calenardhon to Eorl and his people, who until this point had been living in the valleys of Anduin where their growing population had caused them to become 'somewhat straitened in the land of their home' (J. R. R. Tolkien 2008c, 1395). Tolkien crystallizes how land can become a tool of political facilitation: Cirion uses the region of Calenardhon to reward Eorl for his help in the protection of Gondorian land, and thus to reinforce and concretize the alliance between the two peoples. For Eorl and his people, meanwhile, the region of Calenardhon becomes a means of escaping their previously impoverished circumstances and establishing themselves as a newly flourishing state of power within Middle-earth. The land of their home – a land defined not by its potential for political might but by its affective and psychic significance in their people's history – is abandoned. As Glissant noted, such land is defined by an absence of limits, and thus does not allow for the prospect of accretion: the Eorlingas were straitened not only through resources and growing population, but by the afforded possibility of expansion and consolidation of power. It is only by leaving their home, and transitioning into the development of a territory that can be protected, be expanded and form alliances, that the Éothéod can realize their desire for political might. The result, of course, is the kingdom of Rohan, and the military strength of the Rohirrim.

The second model, where the relationship between the land and its inhabitants is driven by emotional and psychological attachment, stands distinct from such exertions of power yet nevertheless remains political, existing as it does within a context of conflict and territorialization. Take, for example, the Shire. As discussed in the second chapter, Tolkien places great stress on the ways in which the relationship between the hobbits and their home constitutes a distinct shift from the hierarchical and exploitative relationships between humans and their environments present elsewhere. Although the land of the Shire technically belongs to the Kingdom of Gondor and is thus part of the same feudal system, its relative political and geographical insignificance leads to dislocation from such centralized power structures, so that it effectively becomes a self-governing territory. The prologue makes explicit that Shire hobbits were 'in name' subjects of the king, but were in practice ostensibly ruled by their own chieftains (J. R. R. Tolkien 2008a, 6), to the extent that the area has 'hardly any "government"' and that families would for the most part 'manag[e] their own affairs' (2008a, 12). Notably, the two official authoritative positions in the Shire's political system – the Thain and the Mayor – are framed as largely titular and ornamental: the Thainship has 'ceased to be more than a nominal dignity', while the Mayor's duties revolve largely, and appropriately for hobbit country, around ceremonial feasts (2008a, 12).

It may be politically decentralized and removed from such structural reinforcements of power, but through the Shire, Tolkien makes explicit the ways in which land is rendered political even beyond 'politics' as a strictly defined system of governance. As is the case with the Tooks and Tookland, certain wealthy families and clans own large areas of land and indeed even have it named after them, the eponym highlighting how the land is made inextricable from its inhabitants and their sense of security and belonging within it. It is, moreover, notable that regardless of whether they have strict administrative possession over it, all hobbits are shown to possess a strong, almost primal connection to their individual part of the Shire. Rather than manifestations of economic or sovereign control, acts of farming, gardening, burrowing or other markedly material forms of shaping and maintaining the land epitomize the deeply felt collapse in boundaries between the hobbits and their homeland. These connections notably transcend class divisions; a landless gardener such as Samwise Gamgee feels as fundamental a sense of attachment to his land as the landed and wealthy Tooks. The exclusively agri- and horticultural nature of the hobbits' activities recalls Fanon's positioning of the land as a source of both bread and dignity: the centrality of land becomes a means of physical and emotional nourishment within the hobbits' formation of an affective relationships with their homeland. Yet this affection, too, results in inevitable territorialization. The Shire is rendered a political entity precisely because of this strictly emotional and psychic relationship it holds for its inhabitants: Frodo envisages its destruction in Galadriel's mirror, spurring him further on his quest, and Saruman – as we shall explore further below – invades the Shire because he well understands the loss it would represent for his enemies. Through the transformation of land into territory, both land and inhabitants are rendered mutually vulnerable, and in response the land is made hostile, militarized through both attack and defence.

The Threshold Space

From the great realms of Gondorian kings and the mighty Rohirrim to the bucolic slopes of the Shire, land in Middle-earth is transformed into a distinctly political entity, one that is both fiercely protected, hungered for and – crucially – explicitly demarcated as a marker of possession. Yet threshold and border spaces that mark out the character and ownership of territory hold significance not merely as an indicator of a realm's political and governmental limits, but of its potential for breaching and incursion. The border is a site of both power and powerlessness. Saïd outlines the material and psychic practicality of borders in creating distinct cultural or political spaces by 'designating in one's mind a familiar space which is "ours" and an unfamiliar space beyond "ours" which is "theirs" …' (2003, 54). Through the political fragmentation of physical land, then, the border both enables the act of invasion – by creating 'other' territories that can be attacked and appropriated – and resists it – by offering a physical and tangible opposition to the assault. Middle-earth is filled with such borders that act both as defensive structures and sketch out the limits of a realm's political power on lands and maps, encapsulating both the enactment and the limits of a realm's power.

This dualistic nature of the borders – as both defensive and potentially offensible – leads to a constant anxiety surrounding these spaces and their effectiveness in warding against threats. When Gandalf counsels Théoden about Saruman's treachery, he claims that it was easy to spy on Rohan, 'for your land was open, and strangers came and went' (2008e, 680), while in the Shire, the number of Bounders (hobbits that patrol the borders) is said to have greatly increased due to the increase in strangers at the Shire boundaries, 'the first sign that all was not quite as it should be' (2008a, 13). Such anxiety immediately positions the border as a tipping point in the act of territorial invasion and control, highlighting its ability to both reject and permit external hostility. This dual role of the threshold space is especially encapsulated in the Gondorian border at Minas Tirith. Pippin's first encounter with the city is its defensive structure: Rammas Echor, the wall that surrounds the Pelennor Fields outside the city. Drawing inspiration from medieval walled cities, Rammas Echor acts as Minas Tirith's main defence and was built following various Gondorian defeats and loss of land. Its determined function is thus and always has been explicitly tied to the protection and maintenance of Gondorian land, and specifically to anxieties surrounding territorial loss. The scale of these anxieties finds its correlate in the immense scale of the wall: it 'loom[s]' out of the mist, has been built 'high and strong', and encircles the city for over ten leagues (2008c, 981). Gandalf and Pippin are challenged as soon as they approach the wall, when a suspicious guard informs them that 'we wish for no strangers in the land at this time' (2008c, 979), positioning Rammas Echor as part of a much broader system of border protection that works to reject Saïd's unfamiliar other. Yet the wall is also distinctly vulnerable: it is 'partly ruinious' and is being hastily repaired as Gandalf and Pippin approach (2008c, 979). The fragile state of the wall, and by extension its inability to act as a successful border, becomes the key factor in

Minas Tirith's easy invasion by Mordor's army; a fact foreshadowed by Gandalf, who warns 'you are over-late in repairing the wall of the Pelennor. Courage will now be your best defence ...' (2008c, 980). For Tom Shippey, the failure of Rammas Echor to act as a defensive structure and its overall ineffectuality within Denethor's poorly strategized defence, find parallels in the French Maginot Line, a failed defensive wall constructed after the First World War along the border between France and Switzerland, Germany and Luxembourg in order to protect against future German hostilities and incursions. The wall's fortifications were weak along the north near the Belgian border – the French government mistakenly believing the terrain would prove too difficult for attack – which allowed the eventual invasion and occupation of France by German forces. Rammas Echor, Shippey argues, is rendered similarly pointless. Denethor's insistence on defending it is depicted as an unequivocal military error: its looming structure does little to quell Sauron's forces and serves only to obstruct the arrival of the Rohirrim and nearly kills Faramir (2005, 170). The historic parallels between the two cement the contradictory character of Rammas Echor: it is both an extensive and ambitious fortification and a failed defensive tool. Its simultaneous characterization – as both high and laid low, strong and fragile, embodying both the frustrated attempts of the border's defence and the effectual attack of the conquering army – positions it, and the border space more broadly, as a site of both attack and defence, force and vulnerability.

However, the physical presence of Rammas Echor forms only one part of the wall's defensive mechanisms. When Gandalf attempts to cross over the wall, the guard on duty acknowledges Gandalf's knowledge of 'the pass-words of the Seven Gates' (2008c, 979) and allows him to go forward. This exchange forms part of a wider pattern in Tolkien's legendarium of the policing of threshold and border spaces through specialized knowledge as well as strength. In Mordor, it is not only the physical impenetrability of the Black Gate and its surrounding walls that frustrate Frodo and Sam's attempts to cross the border, but also the requirement of 'the secret passwords that would open the Morannon' (2008e, 832) and the eyes of the Watchers at Cirith Ungol. At the entrance to the mines of Moria, meanwhile, is another password-controlled border, the solution to which requires an explicit knowledge of Elvish language. And through the ancient Elvish city of Gondolin, Tolkien makes explicit the accretion of both physical and knowledge-based defence in imagining and reinforcing the borders of a place. Like Minas Tirith and Mordor, Gondolin is protected by defensive structures that strikingly accrue through Tolkien's writing process. The earliest draft 'The Fall of Gondolin' only mentions three gates – two that close off the city's outer walls and the main gate – yet in the much later 1951 draft 'Of Tuor and his Coming to Gondolin', the city is now protected by seven named gates: the Gate of Wood, the Gate of Stone, the Gate of Bronze, the Gate of Writhen Iron, the Gate of Silver, the Gate of Gold and the Gate of Steel, which amass not only in number but in strength, value and complexity of material. Yet these are not the city's only fortifications: known also as the Hidden City of Turgon, its positioning within the depths of the Encircling Mountains – a position only known to a few – allows it to resist assault and capture for centuries. The exclusivity of this knowledge is portrayed as key to the city's protection: the Elves refuse to come to the rescue of their comrade Húrin for fear that he will lead enemy forces to them; Tuor – a stranger's – presence is only enabled and legitimized through the aid of the patron god of the city, Ulmo; and the city ultimately falls when one of its own betrays this secret knowledge to Melkor. Gondolin fantastically exaggerates the imaginary of the completely controllable medieval walled city, yet its defence – Tolkien stresses – lies in its knowledge-based inaccessibility as well as sheer, physical impregnability, and when the former fails, so too does the latter.

Through these cities and realms, the consolidation of a land's boundaries – and thus what lies within the boundaries – is demarcated both through the physical and psychic barrier of the border space. The knowledge conditions that Tolkien sets up around these vulnerably conceived spaces works to demarcate binaries of space – between ours and theirs, admissible and inadmissible, naturalized and othered – not only as physical realities, but as a predominantly relational affair: either one belongs through a pre-existing relationship with the land or one is an interloper. It is, at its heart, a Foucauldian system of knowledge-based power, yet its effectiveness lies not just in the accumulation and deployment of knowledge, but in its policing and hoarding. This attention that Tolkien pays to the knowledge of borders as a potential avenue for incursion is emphasized through their mapping. Despite being ostensibly navigational objects, Tolkien's Middle-earth maps rarely illuminate passage through highly protected borders. Minas Tirith and the Black Gate are both portrayed on the Middle-earth map and the map of Rohan, Gondor and Mordor, yet there is no specific path or access indicated, and no mention of the passwords required. Gondolin, meanwhile, is depicted on the Beleriand map, but its depiction only underscores its impenetrability: it is depicted as a small island in a sea of mountains, the narrow white space between its walls and the appropriately named Encircling Mountains only serving to underscore its sheer isolation, rather than signifying an open space through. The exception to this lack of navigational access is, of course, Thror's Map, which both depicts the Lonely Mountain *and* explains how to access its secret entrance through the moon runes on the left-hand side. Yet the use of moon runes – readable only to some, and only under certain celestial conditions – and the careful passing down of the map from Thror to the next heir, limits how this knowledge is distributed. This careful guardedness on the part of the maps, the wary restraint that they exhibit in fulfilling their supposed function – in showing the reader where to go – works to position the border space, and by extension the territories within it, as a site of anxiety and precarity, of constant potential permeability and encroachment.

The borders of lands such as Gondor and Gondolin reflect this anxiety of invasion through their limited distribution of knowledge of the border area; other representations – or marked representational absences – of borders on Middle-earth's cartography meanwhile highlight the vulnerability of the border by depicting it as a politically liminal space susceptible to shifting political control. This is particularly striking when noting the difference between depictions of physical, geographical and topographical borders and those of notional political borders. The borders that tend to be depicted in Tolkien's cartography are those – political or not – which are formed by geographical features or man-made structures: the Lonely Mountain, denoted by its own mountainous walls; the realm of Mordor, clearly separated from Gondor by the Ered Luthui and Ephel Dúath mountains; Gondolin, encased in the valley of Tumladen and the Encircling Mountains; the woods of Lórien, plainly demarcated from the surrounding Drimrill Dale and Field of Celebrant; or Rammas Echor, depicted on both the Middle-earth map and the map of Rohan, Gondor and Mordor as a small, looping line around Minas Tirith. However, the notional political borders that are not fashioned by pre-existing physical formations are notably absent from Middle-earth's maps: their unfixed nature is embodied in their resistance to the fixity of cartographic representation. The Shire is described as having boundaries and borders – made explicit through the Bounders that are said to patrol them[2] – yet these are not marked out on the Middle-earth map, nor are the firm borders between Rohan and Gondor. Doriath in Beleriand, meanwhile, is a realm largely characterized by its border. Doriath itself translates to 'Land

---

[2] The term 'Bounders' almost certainly stems from the early medieval custom of 'beating the bounds', in which members of a parish community would walk the perimeter of the parish so that the younger and newer members could be taught by the elders and church officials where exactly the boundaries lay. The custom is rooted in a pre-cartographical age, where such geographical limits were not communicated through maps, but rather through the oral and ritualistic passing down of knowledge. The Bounders in the Shire mimic such traditions, recalling and reinforcing the relationship between the physical marking out of borders and their absence within corresponding cartographic documents.

of the Fence' or 'Land of the Girdle' in Sindarin (J. R. R. Tolkien 1994, 370), a name which derives from the Girdle of Melian, an enchanted border set around the kingdom by its queen that prevents any strangers from entering the land without King Thingol's consent. The Girdle is said to encompass and protect the Forests of Neldoreth, Region, the West March of Nivrim and the neighbouring area of Aelin-Uial; however, on the Beleriand map, there is no suggestion that these areas form part of the same kingdom, and the (admittedly invisible yet no less notional) Girdle of Melian is nowhere to be seen on the map.

Rather than through fixed depictions of notional political borders, different territories are instead broadly marked out through place names, frequently written in large letters arching over the entire territory, thereby inscribing the politics of the space both onto the map and across the cultural imagination of the entire territory. To a certain extent, this is part of the pseudomedieval stylization of the majority of Middle-earth's cartography: the marking of notional, political borders through cartography is largely a product of modernity and the rigid conceptualization of the nation-state. Nevertheless, the lack of fixed borders within Middle-earth's cartography also acts as a further indication of the political and invasive activity of Middle-earth, and of the border itself as a site of marked vulnerability. Throughout Tolkien's legendarium, the border is frequently breached and the act of invasion is realized. This creates liminal spaces, particularly around borders, which are constantly conquered and occupied by opposing forces and thus slip easily between political territories. Minas Morgul, the fortress that Frodo, Sam and Gollum slip past before climbing the stairs at Cirith Ungol, was once called Minas Ithil and was the twin city to Minas Anor (later Minas Tirith). Established by Isildur and Anarion after the destruction of Númenor, the city was captured by Sauron's forces in the Second Age; shortly afterwards, Isildur recaptured it, and after Sauron's great defeat at the end of the Second Age, the city was re-established as one of Gondor's key settlements and fortresses. Centuries later, in the Third Age, the city was attacked and captured once more after several years of siege, before being subsumed into Mordor and renamed Minas Morgul. After Sauron's final defeat in *The Lord of the Rings*, the city once more reverts to Gondorian rule, and became part of the fiefdom of Ithilien.

The Lonely Mountain is similarly involved in a back-and-forth struggle over land: the dwarves first settled there in the Third Age, but when their hoarded wealth attracts the attention of Smaug, it and its surrounding areas become known as the Desolation of the Dragon. The dragon is slain by Bard, the mountain reverts to the dwarves, although it remains the object of attack several times after: first, by the orcs during the Battle of the Five Armies, later by the Easterlings and finally by Sauron's forces during the War of the Ring. Crucially, Smaug's invasion of the mountain is stressed not as the typical act of imperialism that shifts and pulls at border spaces, but as motivated by pure individualism: Thorin explains to Bilbo that '[t]here were lots of dragons in the North in those days, and gold was probably getting scarce up there ...' (2008b, 31), yet Smaug arrives alone and claims the mountain and its surrounding lands for himself. Tolkien depicts his conquest of the Lonely Mountain not as an organized attack in order to perpetuate a particular political agenda or expand the reach of power of a particular people, but as a means of accumulating self-interest. Smaug interestingly never seeks to expand his dominion: he is drawn to the wealth amassed in the mountain, and his focus remains on this treasure and the land of the Lonely Mountain itself. When he first physically appears in *The Hobbit* he is depicted sleeping on an enormous pile of treasure, which stretches 'about him on all sides ... across the unseen floors ...' (2008b, 273). This visual is underscored by Tolkien's own illustration of Smaug in the original edition of *The Hobbit*, in which Smaug's underside is encrusted with gold and jewels, reinforcing the connection between him and the mountain's wealth by subsuming it into his body. Later, when Smaug guesses that Bilbo came via Lake-town and suspects he was sent by the Lake-Men, he notes that 'I haven't been down that way for an age and an age; but I will soon alter that' (2008b, 284).

The instability of the Lonely Mountain's border space is the realization of anxieties of invasion, so that it is not merely the border space that becomes liminal, but the very land itself as a site of belonging. The violation of both the Dwarves' home and the physical space of the land is reflected in *The Hobbit* maps, namely Thror's Map and the Wilderland map. In both, a drawing of a single dragon is placed prominently over the mountain, and the surrounding area is labelled 'The Desolation of Smaug', both indicating Smaug's presence and stressing the illegitimate nature of his invasion by positioning him as extraneous to the mountain. The presence of the dragon on Thror's Map is also commented on by the dwarves: Balin notes that '[t]here is a dragon marked in red on the Mountain ... but it will be easy enough to find him without that ...' (2008b, 27). Balin's comment may be tongue-in-cheek, but it also draws attention to the purely political nature of the dragon as cartographic symbol: the marking of a dragon on the map will not provide any navigational aid, but rather exists in order to encode the (new) politics of the land within the map. Notably, the dragon was not always present on Thror's Map: in the original sketch of the map, found in an early manuscript of chapter one, there is no dragon; however, a small dragon appears on a later copy of the map 'Copied by B. Baggins', and on all subsequent sketches and drafts, suggesting that its symbolism was a deliberate and eventually integral part of the map. The same process can be seen in the map of the Wilderland: the original sketch – which admittedly lacks much of the iconographic qualities of the final version – is missing a dragon, but it appears in large scale on the final version. As time went on, the inclusion of a dragon symbol became an integral element in telling both the narrative of Smaug's occupation and the long instability of the Lonely Mountain's border space: its vulnerability to acts of material and cartographic (re)inscription.

The colonizing project of the Númenóreans, meanwhile, involves numerous invasions of border areas, notably those which border the sea and act as the main frontier to the landmass of Middle-earth. Tolkien makes explicit the resistance of the native people to these acts of invasion: at times the newcomers 'would suffer great loss and be flung back', implying not just the devastating loss of people but the concomitant loss of land in the constant skirmish that defines the border space (J. R. R. Tolkien 1996, 424). The consequences, both environmental and human – as we shall examine shortly – are enormous. These borders, and notional political borders more broadly, become characterized by their impermanence, where the political instability of the land leads to the instability of the border as marker. The constant movement of Middle-earth's often small, tribal societies further compounds this sense of political volatility: particularly in the early Ages of Middle-earth, the various tribes of Men (including the Drúedain, the Easterlings and the Three Houses of Men), the various migrating tribes of Elves to and from Valinor, the three 'breeds' of hobbit (the Harfoots, Stoors and Fallohides) who dispersed and settled in different areas, or the roving bands of orcs are involved in a near cyclical process of settling regions, being driven out and invading new ones. The absence of political borders on Tolkien's maps becomes an articulation of both the difficulty of establishing and maintaining borders within the physical landscape itself and the broader vulnerability of the border space as a site of precarity, invasion and flux. Throughout his legendarium, the border acts as both a space of reclamation – of delineating what is ours and what is not, of fierce territorialization and fragmentation of land – as well as a marker of the possibility of invasion, of a loss of protection for everyone and everything that called that land home. As we shall examine, such losses have marked material effects: a breach of border space indicates, so often, a consolidation of power built on ecological destruction and the lives (and deaths) of vulnerable people – an entanglement of harm between the human and non-human.

Mutual Vulnerability and Racialization

Such mutual harm is everywhere we look: the dispossession of the Dwarves and the scorched land around the Lonely Mountain after Smaug invades; the wholesale destruction of Beleriand in the struggle for power; the psychic and material damage wrought on both hobbits and Shire during the invasion of Saruman. The latter is a particularly notable example; it contrasts directly with the larger-scale warfare perpetuated by the same offender earlier in the narrative, shifting from the markedly imperialist underpinnings of the attack on Rohan, in which Saruman attempted to expand the reaches of his territory and power, into a far more uncategorizable and wilful destruction. There is an emphasis on the personal: when the hobbits finally confront Saruman after the Battle of Bywater, he makes explicit his reason for targeting the Shire, declaring 'Saruman's home could be all wrecked, and he could be turned out, but no one could touch yours ... one ill turn deserves another' (2008c, 1333) and openly stressing his desire for vengeance in describing the destruction wrought, and how 'it will be pleasant to think of that and set it against my injuries' (2008c, 1333). Although the scale of the occupation of the Shire reaches beyond Saruman, in that numerous other offenders are involved and even perpetuate most of the devastation – both material and human – the confrontation with Saruman – positioned at the end of the chapter, and thus coalescing, concluding and explaining all the episodes of loss and violence caused by the takeover of the Shire – reorients many of the grand narratives of imperialism and power that have formed the thrust of the narrative until then, revealing the ways in which both land and people can become pawns within individual narratives of anger. The smaller-scale nature of the conquest of the Shire is reflected in its disorganized character: the perpetrators are referred to constantly throughout the chapter as 'ruffians', defined by their violent actions while simultaneously underlining their disordered and petty nature. Each attempt by members of the new order to bring Frodo, Sam, Merry and Pippin under control fails spectacularly: Bill Ferny attempts to keep them out of the Shire gates by threatening them, but runs away as soon as he sees their drawn weapons; the Shirriffs – who are not outsiders but nevertheless conform to the mob rule imposed on the Shire and its residents – attempt to arrest the hobbits but their authority is entirely undermined when Frodo laughs at them, Sam answers them back, and the four 'prisoners' proceed at their own pace; and when the hobbits arrive at Bywater, the Men who attempt to stop them flee as soon as they realize the hobbits are prepared to fight back.

Nevertheless, despite the attention Tolkien draws to the ineptitude of its perpetrators, the violence they commit against the Shire remains one of the most shocking of the legendarium, both in how it undoes the safety that the Shire has represented until then and in its totalizing nature. This violence is enacted both against the land and its inhabitants, a prescient embodiment of Saïd's definition of imperialism as an act of 'geographical violence', in which the bringing of physical territorial space under strict control – through exploration, exploitation and administration – is reflected in the way its inhabitants are brought under a new imperialist rule. For Saïd, anti-imperialist resistance – that is, the resistance of the human – must begin with an emancipation of the geographical – that is, the non-human – which has been submitted to such violence (1990, 77). The occupation of the Shire is not an act of imperialism – as discussed above, it lacks the structural hallmarks of imperial warfare – yet even in such small-scale conquest, there is an entanglement between violence against people and violence against land that articulates a sense of mutual vulnerability, of doubled implication. As Frodo, Sam, Merry and Pippin delve deeper into the occupied Shire, they discover both types of damage almost simultaneously. The Shirriffs threaten the four hobbits with the Lockholes, prison cells that are alluded to multiple times throughout the chapter, and which act as a gross subversion of the comfortable, domesticated hobbit holes that are most associated with the Shire. There are also references to more explicit violence: Robin explains that the chief's men no longer stop at imprisonment and 'often they beat' the prisoners (2008c, 1312), and Men jeer at the hobbits to return to their homes 'before you're whipped' (2008c, 1322). This harm to the human is interlaced with damage committed to the Shire's lands: Frodo and Sam's first encounter with Bywater results in their 'first really painful shock', with houses burned down, gardens 'rank with weeds' and ugly new houses and an industrial chimney sprawled through areas of natural beauty (2008c, 1314). There is an emphasis not just on non-human but specifically ecological damage: trees have been felled, the chimney is 'pouring out black smoke', and the 'filth a purpose' now pollutes the Shire's water (2008c, 1314, 1326), literalizing the geographical violence of invasion beyond mere enforced administration of territory. The non-human becomes, rather, as DeLoughrey and Handley express it, a 'participant' in the political processes of the human: the land of the Shire is victimized, both because it will hurt the hobbits, and because it enables the fantasies of industrialization and power that Saruman longed to enact on a larger scale.

For Tolkien, the ecological damage enacted upon the Shire is no less horrific than what is enacted upon its inhabitants, encapsulated in a speech Farmer Cotton makes to the returned travellers:

> All the ruffians do what he says; and what he says is mostly: hack, burn, and ruin; and now it's come to killing. There's no longer even any bad sense in it. They cut down the trees and let 'em lie, they burn houses and build no more.
>
> (2008c, 1325)

Certainly, Farmer Cotton's lament refers to the hobbits murdered during the occupation, their death revealed in two conversations occurring in the chapters preceding this speech, between Mr Butterbur and Gandalf and between the hobbits since they returned to the Shire. However, it is notable that Farmer Cotton immediately follows his comments on needless killing with a reference to the needless felling of trees, trees that are left to lie as embodied spectacles of violence. These killings, both human and environmental, disturb and derange the common order of things. In the case of the murders, Tolkien makes explicit how unnatural the event it is: when Mr Butterbur first refers to them, he exclaims: 'there were some folk killed, killed dead! If you'll believe me' (2008c, 1299), his heightened repetition and tautology disrupting the calm continuity of the Shire even before its ruination is observed. Frodo later builds on the rareness of murder in the Shire, explaining that '[n]o hobbit has ever killed another on purpose ... nobody is to be killed at all, if it can be helped ...' (2008c, 1317). This sense of uncanny inversion follows through in the environmental cost; in the industrialization of a land that was hitherto green and thriving, and particularly in the felling of the party tree that once symbolized the community-minded spirit of the Shire. It is no mere coincidence that the length of *The Lord of the Rings*' narrative is bookended by the party tree: once flourishing as the centre of Bilbo's party, and by the end lying 'lopped and dead' (2008c, 1330). Even after witnessing the wreckage that has spread across the rest of the land, it is only when he catches a glimpse of the party tree that Sam bursts into tears: its loss the 'last straw' (2008c, 1330) in an already overwhelming narrative of loss. Its arc encapsulates the throughline of Tolkien's narrative as an elegy for ecological loss; yet it is striking that this final act of wanton damage, taking place well after the main drive of the narrative is resolved, is bound up so entirely with other incidents of unprecedented (human) death. There is no way, Tolkien tells us, of disentangling power over the human from power over the non-human: there is always a mutual cost.

There is no map of the Shire post-occupation or post-Scouring; indeed, the only drawing that Tolkien made of this point in the story was a floor plan and sketch of Father Cotton's house (Hammond and Scull 2015, 197). Instead, the only cartographic object of the invaded land is 'A Part of the Shire', drawn to illustrate the area before the events of *The Lord of the Rings* transpired and depicting a domesticated landscape that is administratively controlled and well-ordered. It is a map intended, as Stefan Ekman argues, to instil a sense of reliability and order, so that the main elements of the Shire map – topography of vegetation and water courses, road systems, population centres and administrative regions – become 'part of an overarching discourse of defining, situating, and familiarizing the Shire' (2013, 49). This cartographic characteristic works to mirror the depiction of the Shire at the beginning of the narrative as a place of harmony, and extends to other cartographic depictions of the Shire – not provided as visual object but discussed in the narrative – such as Bilbo's map of the Country Round or Frodo's maps, which only show white spaces beyond the Shire borders, boldly demarcating the space between the familiar and the unknown beyond. For Ekman, the map's atemporality, its lack of precise detail that might indicate the kind of sudden change that the Scouring of the Shire precipitates – the party tree, destroyed gardens and knocked over houses that never featured on the map anyway – means that the violence of the occupation does not create any cartographic tension, so that the map exists in the 'constant present' (2013, 51).

And yet, even if the map remains technically accurate, everything that it once represented becomes outdated. What Ekman defines as the map's project of 'defining, situating, and familiarizing' the territory, that is to say, the map's participation in the protected idyll of the Shire, is entirely disrupted. Much like Lothlórien discussed in the previous chapter, a physical place unnaturally protected from the ravages of time, the Shire until its scouring seemingly existed – with no such magical enchantments – beyond the reach of political violence, a bubble of safety that could reflect the unnatural order and fixity that its cartographic correlate depicted. By the end of *The Lord of the Rings*, everything the map culturally symbolized – its safety, its parochiality, its innocuity – has been destroyed, so that the geographical violence wrought upon the Shire takes place threefold: upon its people, upon its land and upon representation of its land, overturning the certainty and familiarity which the map originally promised and transfiguring the Shire into a burnt and maimed shadow of its former self. Frodo and his companions may drive evil out of its lands and Sam may replant the trees to magnificent effect, but evil was allowed to take place and, as Frodo says much later, some things never fully heal. Here, in the supposed idyll of the Shire, the totality of violence – its human and non-human reach – has been made evident.

The events of the Scouring of the Shire reveal the ways in which Tolkien was attuned to the doubled violence of invasion, the destruction that is wrought both upon a land and its people. Saruman's attention on the Shire, motivated by revenge as it is, is a racial one in the strictest sense of the term; his plans for power having been thwarted by hobbits, Saruman sets his sights upon their homeland, extrapolating his hatred towards the hobbits of the Fellowship into a hatred, and enacted harm, towards their people. Yet there is no real racialization of the hobbits at play in the narrative, either diegetically or extradiegetically: the hobbits are not targeted because they are thought of as inferior (or, at any rate, no more inferior than the Men or other mortal beings that Saruman also sets himself against) but because they, as a people, enacted a resistance towards his grab for power that Saruman finds intolerable.

Yet in reality, most acts of power over the land involve, or justify themselves through, a racialization of the colonized subject; as Huggan and Tiffin argue above, the process of colonization can only be enacted over the 'not-human', that is to say, what has been rendered non-human and inhuman in turn. Tolkien's relationship with colonialism and race, and his concomitant engagement with it in his legendarium, is – as discussed briefly in the Introduction – a thorny and often contradictory one. On the one hand, there exists a significant body of personal writing that makes explicit his disavowal of imperialist processes and the particular mechanisms of the British Empire under which he was living. In a letter to Christopher in 1943, Tolkien laments the globalization of American and British culture and voices his concerns that a victory for the Allies at the end of the Second World War would not necessarily be an unproblematic result, before defending himself against letter censors by affirming his patriotism: 'I love England (not Great Britain and certainly not the British Commonwealth (gr!))' (2006, 65). In a 1945 letter, Tolkien again worries about the implications of the end of the war and the weapons – such as Christopher's RAF planes – that are used to bring it about, explaining, 'I know nothing about British or American imperialism in the Far East that does not fill me with regret and disgust ...' (2006, 115). Notably, Tolkien's lack of enthusiasm was not only reserved for the twentieth-century British and American imperialism, but for the structuring power of empires more broadly; in a 1944 letter, he remarks that, if he were a citizen of the Roman Empire, he would have been unable to hate the Gauls and Carthaginians out of pure patriotic allegiance, adding that 'I should have hated the Roman Empire in its day (as I do)' (2006, 89). This dislike of empire results throughout his legendarium, as we shall shortly examine, in a marked sympathy for the othered colonial subject, and an often surprisingly incisive grasping of the human cost of the racialization that enables colonialism.

On the other hand, and no less important, is the overt anti-Black, anti-Asian and anti-Middle Eastern racialization that is fundamental to Tolkien's worldbuilding structures. Setting aside the whiteness of the protagonists due to the text's Eurocentric focus, many of the legendarium's malevolent characters are strongly racialized beyond frameworks used to unpack colonial and imperial power structures. Margaret Sinex draws attention to the strong Middle Eastern racialization of the Easterlings, described repeatedly through *The Lords of the Rings* as 'dark' and 'swarthy', arguing that such conceptualizations mimic and proliferate the marginalization and vilification of Saracen men in medieval Christian imagery. The orcs, famously, are described using similarly racialized terms, and in a 1958 letter to Forrest J. Ackerman, Tolkien explains that their physical features were inspired by 'degraded and repulsive version of the (to Europeans) least lovely Mongol types' (2006, 274). Tolkien's parentheses in the letter demonstrate his awareness of white European racial bias, yet his decision to take advantage of and perpetuate rather than challenge this bias remains deeply troubling.

To ignore either aspect of Tolkien's engagement with race would be to neglect either the at-times complex anti-colonial threads that run through his work or to write off the unacceptable racialized stereotypes that prop up the legendarium's narratives around power and act as often lazy shortcuts towards supposedly fixed ideas of good and evil. As Elizabeth Massa Hoiem argues in her analysis of colonialism in the legendarium and particularly in 'Aldarion and Erendis: The Mariner's Wife', Tolkien offers a 'sophisticated criticism' of colonialism throughout his legendarium while nevertheless 'making use of the colonial rhetoric that saturated the literature of [his] time' (2005, 76). The imagery and language to which Tolkien resorts remain rooted in the contemporary racial power dynamics of the early and mid-twentieth century even as they attempt to confront them, drawing attention to the intersection between racism, colonization and land without fully subverting the cultural frameworks that uphold this. To read Tolkien is to have to hold both truths in mind at the same time: the ways in which he understood how people are made vulnerable alongside their land, the ways in which he failed to grasp how the maintaining of racial narratives play an enormous part in the enablement of such power.

_The contradictions of Tolkien's approach are held in the history of the Wild Men of the Woods, or the Drúedain, which unravels as an alternative and distinctly more tragic narrative of loss of home in comparison to the invasion of the Shire: a portent of what might have occurred if the agency and power that the hobbits were granted, both on a narrative and political level through their alliances with Middle-earth's more powerful forces, were denied. Modelled on the medieval character of the 'Wild Man', an 'archetypal outsider' (Flieger 2003, 95) who exists on the boundaries of the civilized world, the Drúedain nevertheless eschew many of the typical attributes of the medieval Wild Man: they are capable of speech and thought, and align themselves with the side of 'good' in the battle between the Free Peoples of Middle-earth and Sauron, despite their abuse at the hands of the Men of Númenor and Rohan over the years. The numerous similarities between the Drúedain and hobbits only serve to emphasize the former's more beleaguered path through a very comparable narrative. Like hobbits, the Drúedain are described as a diminutive race that are largely unknown and ignored in the wider history of Middle-earth. However, unlike the Drúedain, hobbits are firmly established and secure in their homeland: they are allowed to live freely and prosper until the events in *The Lord of the Rings*, their attachment to their land is signified through the mapping of it, and even when their land is invaded by hostile forces, they nevertheless succeed in winning it back in the final act of heroic battle in the novel, a geographical and concomitant sociopolitical liberation that recalls Saïd's argument that the first step in anti-imperialist (or, in this case, anti-invasive) resistance is the emancipation of the geographical. The Drúedain, meanwhile, suffer the same predicament as the hobbits in their experience of invasion, yet their fate is very different. Even when they first arrive in Middle-earth and settle in the White Mountains, they are described as 'suspicious of other kinds of Men by whom they had been harried and persecuted as long as they could remember …' (1991, 383). This persecution does not end in the establishing of a new homeland: they are driven out first by tall Men from the East, then by the Númenórean invasions, and then by the Men from Gondor and the Rohirrim, so that eventually only a small fraction of their populace remains. In his description of the Drúedain's displacement, Tolkien uses an iteration of the word 'survivor' three times in as many sentences, drawing attention not only to the remaining Wild Men but also implicitly to all those who have been killed.

The erasing of the Wild Men from their homes, and indeed from Middle-earth itself, is mirrored in their cartographic representation, or lack thereof. There is no indication of the settlement of Drúedain who live in the White Mountains on the Middle-earth map that appeared in the initial publication of *The Lord of the Rings*, nor of the enclave who live north of Anfallas in a land called

Drúwaith Iaur, which literally translates to Old Pukel Land.³ Moreover, there are no individual maps dedicated to these specific regions. These cartographic lacunae both indicate the instability of the Drúedain's homeland, continually shifting due to their displacement, and also emphasize the historically unvalued and thus undocumented fact of their existence on Middle-earth's lands. Rather, the key Drúedain settlements of the Third Age are subsumed within the broader kingdoms of Rohan and Gondor, as represented by the arching letters labelling and claiming the land, inscribing the illegitimate colonization and displacement of the Drúedain on a cartographic as well as a geographical level. This systemic erasure recalls and compounds the function of maps as tools of power; the 'silences' that Harley urges need examination when deconstructing a map (2001, 153) are legible here, a cartographic absence working to articulate the systems of power that generate historic and political discourse in Middle-earth. The map, and by extension the map-maker, create and thereby control the narrative, so that the Drúedain, who are considered an inferior race by those who have access to cartographic production, are erased from both the map and the land.

Why do the Drúedain suffer such continual displacement and persecution, when similarly placed people such as the hobbits escape relatively unscathed? The answer lies in the fundamentally racialized character of the Drúedain, a people drawn in terms that would never be used for the distinctly English-coded hobbits. Their physical features are explicitly othered throughout the narrative: Merry's first impression of Ghân-Buri-Ghân, the headsman of the tribe, is of a 'strange squat shape … short-legged and fat-armed, thick and stumpy, and clad only with grass around his waist' (2008c, 1088), while in *Unfinished Tales*, they are described as 'unlovely', with 'wide faces … deep-set eyes with heavy brows, and flat noses' (1991, 377). The latter account is particularly revealing, intended not as a portrait of an individual but rather an anthropological account of an entire people. There is no escaping the overtly negative nature of such descriptions, language used to craft both a deeply unflattering and explicitly primitive portrait in both appearance and garb. This primitiveness is further emphasized in the harsh and ungrammatical speech of Ghân-Buri-Ghân, which Flieger comments is akin to a 'Hollywood Tarzan' (2003, 100). Such anxieties around the Other, negotiated through a denigration of appearance and intelligence, place the

---

³ Drúwaith Iaur is added in the revision of the map made by Christopher Tolkien entitled 'The West of Middle-earth at the End of the Third Age'. The Drúadan Forest is also depicted in the more topographically accurate map of Rohan, Gondor and Mordor; however, I would argue that the absence of both the forest and Drúwaith Iaur from the original map intended to give a comprehensive representation of the entirety of Middle-earth is significant.

accounts of the Drúedain firmly within Tolkien's contemporary context, calling back to the prevalence of such writing in nineteenth- and early twentieth-century imperialist literature that similarly use unattractive physical descriptors in order to create a racial hierarchy. In *King Solomon's Mines*, H. Rider Haggard describes the King of Kukuanaland with 'lips ... as thick as a Negro's' and a 'flat' nose (1998, 141); the villain of Arthur Conan Doyle's *The Sign of the Four* (1890) comes from the Indian Andaman Islands, whose inhabitants are described as 'naturally hideous, having large, misshapen heads, small fierce eyes and distorted features' (1980, 144); in 'The Adventure of the Three Gables' (1927), a former black slave visits Sherlock Holmes, who stares at 'the visitor's hideous mouth' (1974, 86); and in *Confessions of an English Opium Eater* (1821), Thomas de Quincey encounters a Chinese man with 'sallow and bilious skin ... small, fierce, restless eyes ... [and] ... thin lips' who later haunts his opium-induced dreams (1967, 203).

The characterization of the Drúedain in such similar terms undoubtedly owes much to this tradition, and certainly correlates with the problematic racial hierarchy that is at work throughout Tolkien's legendarium. Yet, as we have examined, there are strange contradictions at work in Tolkien's examinations of race. Notably, Tolkien's depiction of the plight of the Drúedain as a displaced people, if not the physicality of the Drúedain themselves, is largely sympathetic, contrasting heavily with the indifferent portrayal of non-white races in the texts above: the Drúedain are 'harried and persecuted' by other kinds of Men, and the Men of Rohan do not recognize the Drúedain's 'humanity' but instead 'hunt them like beasts', the racialization that has long dehumanized them suddenly depicted as the cruelty that it is (*Return* 1090, *Unfinished Tales* 384). When Éomer questions Ghân-Buri-Ghân's strategic knowledge of Sauron's plans, meanwhile, Ghân-Buri-Ghân's angry reply responds to an entire history of oppression: 'Wild men are wild, free, but not children' (2008c, 1090, 1991, 384), he retorts, before proceeding to lead the armies successfully to their goal. Could Tolkien's use of language possibly be a mere rhetorical tool, using the same linguistic imagery as was used for non-Caucasian races in popular imperialist literature in order to highlight the implementation of intersections of power between dominant and historically marginalized races within his world? This is perhaps giving intertextuality too much credit, particularly when considered in the light of the broader language around race that permeates the legendarium. Yet there is certainly a curious acknowledgement of the injustice that the Drúedain are forced to suffer, and the horrors that have been facilitated through inhuman characterization, at play here. For Dimitra Fimi, the characterization of the

Drúedain bears distinct similarities to eighteenth-century romanticized ideals of the 'noble savage', 'a primitive man who is free, peaceful, and close to nature' (2009, 150). Her comparison calls attention to the colonial foundations that underpin the narratives of the Drúedain and those of displaced indigenous people to whom the concept of the 'noble savage' typically referred, and lays bare both the thrust and inadequacies of Tolkien's anti-colonial critique: the ways in which he was troubled by narratives of racialization, the ways in which he could not fully engage with the realities of historically racialized people as fully realized humans.

The construction of the Drúedain as racially inferior, and the implementation of this hierarchy to both deprive them of and erase them from their land through geographical violence and cartographic absence, follow distinct patterns of settler-colonialism, and the dehumanization of a people in order to take control of the broadly non-/inhuman. Perhaps Tolkien's most rigorous anti-colonial critique in this narrative is his positioning of the Drúedain as the land's rightful inhabitants, through a relationship that transcends those of its outside invaders. We see this in the Drúedain's guiding of the Rohirrim to the Battle of the Pelennor Fields: Ghân-Buri-Ghân explains that the wain-road through the Drúadan Forest has been largely abandoned and forgotten by the Men of Rohan and Gondor, 'but not by Wild Men' (2008c, 1089). As Ekman argues, this episode demonstrates an inarguably more authentic relationship with the land: 'they have lived in this area longer than the people of both Gondor and Rohan ... who are now its masters' and have maintained their knowledge and care of the land, unlike the 'High Men' who have forgotten it in favour of their 'Stone-houses' (2013, 136–37). This innate knowledge counteracts the long work of colonial and cartographic erasure that attempted to dislocate the Drúedain from the map: it is a knowledge that functions in lieu of, and in resistance to, the necessity of cartography. When Aragorn ascends to Gondor's throne and proclaims the forest is to be returned to the Drúedain, with no man permitted 'to enter it without leave' it is – despite the frustration that a white man, one who is the product of long histories of settler-colonial powers in the land, is the person to authorize this reclamation – one of the most unequivocal indications that a more just era is being ushered in.

Narratives of Imperialism

The examples of exercised power over land discussed until this point bear many of the hallmarks of postcolonial theorizations regarding the doubled violence over the human and non-human that occurs in acts of colonial and imperial warfare. Yet, with the exception perhaps of the Drúedain, whose persecution and displacement clearly mimics existing models of settler-colonialism, many of these examples lie beyond such explicit structures of power: Saruman's invasion of the Shire or Smaug's settlement of the Lonely Mountain operate on a small scale, while still demonstrating the entanglement between violence – any kind of violence, imperial or not – over land and inhabitants that occurs. Yet the legendarium does feature explicitly imperial acts that largely reflect Tolkien's own views on the imperialist project at large. There are two such movements in Middle-earth's history: that of the Númenóreans, who sought to expand their empire beyond the boundaries of their island, and that of Sauron and Saruman during the events of *The Lord of the Rings*, who both used military force to gain land and build a geographical empire while also exercising the overlap of human and geographical violence through an exploitation of natural resources and the subjugation of native inhabitants of conquered lands. The key distinction between the two lies predominantly in their method of empire-building; while the Númenórean invasion and colonization of Middle-earth is reminiscent of examples of Western imperialism post-Columbus, journeying over sea to conquer 'unknown' lands, Sauron and Saruman build military empires across the land itself, invading and occupying known and certainly inhabited territories.

This tension between the unknown and known is a key mechanism within the imperialist imagination, which has historically used the pretext of perceived blank space as justification for the seizure of land – the land's unknown quality becoming, rather than actual fact, a deliberate construction in order to facilitate the ongoing imperialist project. In her study on the histories of imperialism, Barbara Bush points to the contemporary conceptualization of the Americas, the Pacific Islands and other sites during European colonization as 'terra nullius' – that is to say, virgin territory – as an important justification of the imperialist project: 'the world unknown to the Europeans was there to be conquered' (2014, 14), Bush explains, so that any unknown land became synonymous with a state of emptiness and availability. This imagination of desired land as both unknown and uninhabited had its equivalencies in cartography. Harley argues that the entanglement between colonialism and mapping, and in particular the use of maps to claim land cartographically before the act of physical language, worked as a form of 'colonial promotion', so that effectively 'maps anticipated empire' (1988, 282). Harley further draws on D. W. Meinig's analysis of the partitioning of North America, who argues that the 'very lines on the map exhibited this imperial power and process because they had been imposed on the continent with little reference to indigenous peoples, and indeed in many places with little reference to the land itself' (qtd in Harley 1988, 282), a practice founded on the imperial imaginary of conquered land as effectively blank and open to both material and human population. This imaginary of 'vacant' spaces found its correlate in imperial literature as well as cartography: in one of the most famous discussions of imperialism and mapping in literature, found in Joseph Conrad's *Heart of Darkness* (1899), the narrator Marlow describes the 'blank spaces' on maps which would draw him in as a child, and how, as the world became explored and colonized, they 'ceased to be a blank space of delightful mystery – a white patch for a boy to dream gloriously over …' (1971, 12). Crucially for Marlow, the realization of the map – and its concomitant land – as not blank at all does not arise from an acknowledgement of its always unblankness, and the settlements and people who have long inhabited its lands, but from the relentless machinations of the imperialist process that fill in what was already there.

Strikingly, Tolkien abstains from the Conradian model of land as explicitly blank space, instead grappling with what has always underlain this myth: the fact of the imperialist process as an act of writing over an already inscribed land. When Frodo slips on the Ring after escaping from a desperate Boromir, the whole of Middle-earth comes into his view, including 'wide uncharted lands, nameless plains, and forests unexplored' to the east (2008a, 522). Yet, while the east is largely uncharted, it is strikingly never the site of imperialist invasion in the legendarium; indeed, it is from the east that armies come to invade the meticulously charted west. In another instance, Frodo looks through maps of the Shire and 'wondered what lay beyond their edges', noting that they 'showed mostly white space beyond its borders' (2008a, 57). Rather than an incentive to exploration and expansion, however, this characteristic of hobbit cartography is depicted as an example of the Shire's parochial myopia; there is never any move on the part of the Shire to investigate and settle these blank spaces. Rather than a filling in of these blank areas, imperialist mapping in Middle-earth is an exercise of power over what areas of land are already explicitly filled in. Borders and the liminal spaces surrounding them are written over by the imperialist act, both figuratively and literally: on the Middle-earth map, the previously Gondorian city of Minis Ithil has been rewritten as Minas Morgul to reflect its conquest by the forces of Mordor and the enormous physical and political changes that the city has undergone. The map acts, if not as a literal palimpsest, then as a figurative one; in areas such as Carn Dûm, where it specifies '[h]ere was of old the Witch-realm of Angmar', the land's previous political allegiance still peers through.

Imperialism becomes an act of erasure: cartographically, geographically and figuratively. We see this in the recurrent imagery used to describe Sauron and Saruman's empire throughout *The Lord of the Rings* as a physical suffocation of previous lands, territories and peoples imagined through fire and smoke. When Gandalf first tells Frodo about Sauron's ambitions to regain the Ring, he warns him that Sauron will 'cover all the lands in a second darkness' (2008a, 67); later at Rivendell he reminds Pippin that the true Lord of the Ring is 'the master of the Dark Tower of Mordor, whose power is again stretching out over the world …' (2008a, 294); Glóin tells the Council of Elrond that 'the Shadow grows and draws nearer' (2008a, 314); when Gandalf is imprisoned at Isengard, he notices that 'whereas it had once been green and fair, it was now filled with pits and forges … Overall his works a dark smoke hung and wrapped itself about the sides of Orthanc …' (2008a, 339); and later Treebeard emphasizes to Merry and Pippin that 'down on the borders they are felling trees – good trees … most are hewn up and carried off to feed the fires of Orthanc. There is always a smoke rising from Isengard these days …' (2008e, 617). Such imagery of shadow and darkness visually reinforces the overlying processes of empire-building; although these do not yet have chartable effects, they act, in a strange subversion of Harley, as an anticipation of the material and cartographic power that has already been demonstrated in Minas Ithil. They also draw attention to the various potentials of violence that such acts of overwriting have: the human costs that will emerge from Sauron's violent expansion, the non-human costs of the urgency towards industry and power that the smog foretells.

The conceptualization of imperialism as an act of erasure and writing over is embodied in the story of Tal-Elmar, a short tale detailing the establishment of Númenórean dominions in Middle-earth from the perspective of the Middle-earth inhabitants who already live in these lands, in particular a young man Tal-Elmar who is caught by a Númenórean landing party. The positionality of the story is key: told entirely through the eyes of the colonized, it reverses the typical privileging of the Númenórean perspective that runs throughout Tolkien's legendarium, from the tales of Númenor themselves to the concerns of its legacy from the likes of Aragorn and Elrond. This shift in viewpoint allows Tolkien to frame the conquest of land that takes place as an act of explicit and unjustifiable colonial violence, rather than framing it through the impulses towards exploration or expansion that thread through other histories of Númenor, such as 'Aldarion and Erendis: The Mariner's Wife'. In Tal-Elmar, the lands on Middle-earth are established as vulnerable to colonizing attacks: 'hateful and proud' Men from the East come with 'bright weapons … thrusting towards the Shores of the

Sea, driving before them the ancient dwellers of these lands …' (1996, 424). It quickly becomes clear, however, that this is not the only experience of violence that these lands and their people have undergone; Tal-Elmar's father Hazad spies ships on the horizon and explains that while the Men from the East are indeed to be feared, the ships herald the arrival of the Númenóreans, who have become legendary for their violent settlement of the lands. These conquests are explicitly imperialist in nature, forming part of a wider deliberate occupation of Middle-earth's lands by the Númenóreans, who – in contrast to the, if not heroic, then distinctly noble and ancient character they are normally depicted through – are depicted as merciless and cruel in their desire for land. It is in this tale that Tolkien makes clear the ugly realities of empire-building: the Númenóreans are described as worshippers of Death who 'slay men cruelly in honour of the Dark' (1996, 427); when Tal-Elmar comes close to them he is frightened, as 'the tales of the "blades" of the Cruel Men were familiar to his childhood' (1996, 433); he is surrounded by 'armed men' (1996, 435); and the Númenóreans bid the native inhabitants to leave 'or be slain' (1996, 437). Moreover, when the Númenóreans first land, they set up lookouts over the land immediately; Tal-Elmar sees them 'on watch; every now and then he caught a flash as some weapons or arms moved in the sun' (1996, 435). The act of surveillance becomes bound up with the more explicit violence of the Númenórean's overt displays of militarization, so that both land and people are brought under the threat of control and death.

This doubled violence against land and people that underlies all imperialism is made explicit throughout: the narrator baldly states that '[t]he object of the Númenóreans is to occupy this land, and in alliance with the "Cruels" of the North to drive out the Dark People' (1996, 436), while further on the Númenóreans tell Tal-Elmar, '[y]our time of dwelling in these hills is come to an end. Here the men of the West have resolved to make their homes, and the folk of the dark must depart – or be slain …' (1996, 437). The unique narrative perspective of the tale – told through the eyes of those experiencing the effects of Númenórean imperialism, rather than the typical Númenórean privileging – allows the act of imperialism to be mediated through the victims' affective response, rather than through simple narrative recounting. When Hazad first glimpses the ships he is 'troubled', and speaks of the 'dread' and 'shadow of fear' that the Númenórean invasions bring to the native inhabitants (1996, 426, 427). Later when Tal-Elmar is forced to spy out the landings, he is 'shaken with fear', 'trembled', is 'afraid', 'quailed' and feels 'terror' throughout his encounter with the invaders (1996, 433, 434, 436). This affective rendering shifts the narrative complexity of the story away from the Númenóreans, their relationship with power and their ongoing project to stave off their fear of death and onto those who linger on the edges of the great Middle-earth narratives, and who bear the brunt of these sweeping desires for grandeur. Here, the looming figure of the Númenórean ships on the horizon becomes a portent of a barbaric kind of violence rather than any kind of 'adventure' of empire-building. Imperialism becomes an act of both geographical and existential erasure that writes over both the sovereignty of a land's native people and their humanity and subjectivity.

The Tal-Elmar tale also crucially demonstrates the intersection between imperialism and other discourses of power. The story explicitly tackles race, although like the history of the Drúedain and other interventions that Tolkien makes into race, his consideration of these power dynamics and in particular the language he uses to illustrate them is not unproblematic. The Númenóreans and the tribes of Men from the East are white and the colonized people are dark-skinned, while Tal-Elmar's grandmother Elmar comes from the white Men of the East, and is captured after a battle and married to Buldar, Tal-Elmar's grandfather. When describing Tal-Elmar's family who take after his grandfather's side – that is to say, the native peoples of the land – Tolkien describes them as 'broad, swarthy, short, tough, harsh-tongued, heavy-handed, and quick to violence' (1996, 423). Much like the Drúedain, their physical appearance is sublimated to the tallness, whiteness and concomitant elegance of white Men and Elves; Tolkien swiftly transfigures these 'inferior' physical attributes into characteristics that align with racialized assumptions made about 'uncivilized' or 'primitive' peoples, namely, that they lack rationality, communication and the capacity for peace. This is made explicit later when Tal-Elmar confronts the Númenórean invaders, where he recognizes their kinship despite being 'born and bred in a decaying, half-savage people' (1996, 434). Here, although the perspective rests with Tal-Elmar, it is the narrator that makes this claim, so that Tolkien's narrative voice remains embedded in and inextricable from the very colonial voice he examines in the story.

The tale of Tal-Elmar therefore cannot be read as a complete rejection of the white supremacist ideologies that have historically enabled Western imperialism, and to do so would be to carelessly disregard the ways in which racist language and systemic prejudice create, maintain and execute harmful power dynamics. Yet, although Tolkien does uphold – as Hoiem terms it – the 'colonial rhetoric' of his time (2005, 76), the tale of Tal-Elmar does nevertheless engage with the ways in which the very frameworks that Tolkien writes within contribute to the imperialist project that he critiques. Tal-Elmar's family history is briefly given – decades previously, his grandfather went to war and brought back with him a Númenórean woman named Buldar, whom he married against her will. In a confrontation between Elmar and Buldar, Elmar laments being held captive among a people that she terms 'base and unlovely' (1996, 425). Buldar responds: 'Base and unlovely thou namest us. Truly, maybe. Yet true is it also that thy folk are cruel, and lawless, and the friends of demons. Thieves are they. For our lands are ours from of old, which they would wrest from us with their bitter blades. White skins and bright eyes are no warrant for such deeds ...' (1996, 425). It is perhaps the most surprising speech within the entire legendarium, entirely destabilizing the impunity with which the narratively privileged races in the legendarium treat those – dark, primitive, lawless – they consider undeserving of interiority and complexity. In this speech, Buldar – and Tolkien through him – gets to the heart of how racism informs imperialism: the false tool of racial purity as a justification for what amounts to simple thievery, and the both ethical and legal injustice of such acts. The specific mechanisms of environmental racism are at play here: the ways in which Tal-Elmar's people are racialized as primitive and animalistic mimics the characterization of the natural world as wild and uncontrolled; indeed, in the Tal-Elmar tale, the Númenóreans' disdain for the native peoples aligns with their summation of the land as 'accursed' and 'dark' (1996, 435), recalling the famous rendering of Africa as abject in Conrad's *Heart of Darkness*.

Huggan and Tiffin's examination on the symbiotic relationship between the oppression of people and the oppression of land provides an illumination on the ways in which the Númenórean empire endangers both. Not only are the native people racialized in such a way as to deprive them violently of their land, but the land too is simultaneously rendered inferior and dispensable. This has been discussed at length in the second chapter, but it is worth revisiting briefly here to consider the explicit connections between imperial violence and environmental damage. When the Númenóreans first begin establishing ports at Lond Daer, the native inhabitants only become hostile when the deforestation becomes 'devastating' (1991, 263), an affective framing that highlights both the ecological and spiritual effects of the loss of this homeland. In response, the Númenóreans become 'ruthless' (1991, 262), treating the native inhabitants as their enemies and felling trees with no thought of husbandry or replanting. There is a disturbing causality here between the Númenóreans' hostility to the people of Lond Daer and their subsequent treatment of the environment: the land becomes caught up in human struggles, and is damaged in order to provoke or injure the humans who live on it.

Although Tolkien focuses on the ways in which the natural and human worlds are enmeshed in imperialist violence, he also takes care to focus on what DeLoughrey and Handley term the 'biophysical' effects of colonialism and imperialism, which include pollution, deforestation, desertification and other tangible impacts outside anthropocentric concerns. Gerard Hynes traces the increasing entanglement of imperialism and deforestation and environmental damage through successive drafts of the Númenor tale. In the earliest version of the story, 'The Fall of Númenor', Númenor only seeks to invade and master the immortal lands of Valinor; Hynes notes that the Númenóreans 'tarried not long yet in Middle-earth, for their hearts hungered ever westward for the undying bliss of Valinor ...' (qtd in Hynes 2013, 125). By the time of 'The Lost Road', however, their ambitions have increased: Herendil says to his father Elendil that Númenóreans desire 'to set foot in the far West, and not withdraw it. To conquer new realms for our race, and ease the pressure of this peopled island, where every road is trodden hard and every tree and grass-blade counted ...' (1987, 60). Although at this stage of the legendarium, these imperial desires are depicted as a product of Sauron's malign influence, Tolkien continues to embed them deeper into Númenórean culture and history, so that by the time of the Appendices to *The Lord of the Rings*, Númenor's imperial expansion has been pushed back to the year 1200 of the Second Age, roughly 2,000 years before Sauron encounters Ar-Pharazôn and tempts him into invading Valinor.

Alongside this rising engagement with imperialism, Hynes notes an increasing awareness of its environmental impact: the Númenóreans deforest their own island and the lands of Middle-earth at a 'devastating' rate. The specific purposes of this deforestation reinforce the harmful link between imperial conquest and the environment and the ways in which the environment is rendered – as DeLoughrey and Handley argue – a key victim rather than mere collateral damage of the imperial project. The Númenóreans cut down swathes of forests, both at home and in occupied land, in order to harvest timber for their ships, so that the trees and their wood directly facilitate the further expansion of the empire that seeks to damage them. Aldarion, who – as discussed in the second chapter – precipitates this deforestation, cares nothing for the trees for their own sake: in an argument with his father after one of his voyages where he returns with two new ships built, and further ships laden with timber, he claims, '[t]he work of forestry I took up, and I have been prudent in it; there will be more timber in Númenor ere my day ends than there is under your sceptre ...' (1991, 180). There is a slippage in Aldarion's language that betrays his singular, myopic anthropocentric concerns: he refers to trees as timber, and deforestation as forestry, revealing not only his utilitarian approach to the natural world, but a perspective on the world entirely shaped, and indeed blinded, by imperialist ambition.

Much as imperial mapping seeks to rewrite previous lands and boundaries, this imperialist drive also rewrites the very landscape of the natural world. The racial violence that undergirds the Númenóreans' actions, which allows them to act with impunity towards the racialized natives of Middle-earth, becomes, as Christine Sharpe argues, the weather: it produces new ecologies. The Númenóreans 'denud[e]' the lands of Lond Daer, driving 'great tracks and roads into the forests northwards and southwards from the Gwathló …' (1991, 262), while in Númenor itself, the land vacillates between being deforested with no thought of replanting, to 'new woods set to grow where there was room' (1991, 190), according to Aldarion's mood and the state of his relationship with Erendis. Crucially, this connection between imperialism and degradation – both the environmental degradation and the moral degradation of the Númenóreans – is laid firmly at the feet of the Númenóreans. Hoiem argues that

> [u]nlike the familiar *Heart of Darkness* degeneration motif, corruption is refreshingly unconnected to contact with colonial Other. Decline directly results from a cultural shift in how we assign value to things. In Tolkien's tales, colonization inevitably commodifies personal and natural resources and justifies questionable actions in pursuit of the dream of progress.
>
> (2005, 77)

The Númenóreans' inability to assign value to the natural world for its own sake and the ensuing violence with which they treat it leads to their inevitable moral and social decline. As Hoiem notes, rather than correlating this corruption with an encounter with the primitive Other, Tolkien frames it as a product of the Númenóreans' unquenchable desire for power in order to locate the blame within the imperialist project itself, and to demonstrate how this practice victimizes both the land and its inhabitants. Although there are no explicit maps of the Númenórean empire, the 'Description of the Island of Númenor' explains that there existed many maps in the Númenórean archive as 'many natural histories and geographies were composed by learned Men in Númenor …' (*Unfinished Tales* 165), which suggests a culture of mapping the island and the broader empire, even if none are provided paratextually; there are also maps, namely the maps of Beleriand, Númenor and Middle-earth, that when considered together represent the profound effects of the empire. Ultimately, everything that occurs in the various Númenórean legends – the racial violence visited about Tal-Elmar and his people, the vast deforestation of Middle-earth that Aldarion enforces to the horror of his wife – culminates in the destruction of Númenor, the realization – and downfall – of the Númenóreans' ceaseless quest for power and immortality. There is perhaps no greater act of imperialist overwriting than the complete reshaping of Middle-earth's shorelines in the wake of the Akallebêth and the globing of the Earth: the island of Tolfalas isolated and alone, the land entirely deforested.

It is impossible to separate out our own current ecological crisis from the history of racial violence that has shaped, both metaphorically and literally, the conditions of our world; it is equally impossible to ignore the violence over the human that results in the long cycles of non-human cataclysm and harm that recur again and again in Tolkien's legendarium. Throughout the formidable length of his narrative, the indifferent ecology of the land is politicized and territorialized, marked out by borders and cultures and administration, prized both by its inhabitants and those that wish to claim it. Anxieties around one's place in the land thread throughout Tolkien's Middle-earth writings, as peoples are displaced and re-established, and as the land becomes a theatre to the violence and extraction unleashed both between humans and over the non-human. Tolkien's distaste for the ravages of empire, colonization and the wielding of power over land is discernible throughout his writing: in the sympathy he extends to the clearly brutalized Drúedain, to the reversing of narrative positionality in the story of Tal-Elmar and the '[w]hite skins and bright eyes' that act as impunity for acts of barbarism and warfare. And yet, as demonstrated, Tolkien falls short of the radical critique he proffers: the legendarium remains rife with instances of colonial thought and rhetoric that embed and stabilize the strength of such institutions, that undo much of the understanding of environmental racism he moves towards.

What does this tell us about Tolkien as a writer? This, for me, is by far the most interesting question. For too long, the conversation around these matters has dwelled on what this tells us about Tolkien as a man, what his often-troubling attitude says about the politics of one of the genre's most beloved literary figures and, by extension, those who are attached to his work. But to critically engage with the work of a writer is to divorce one's affective relationship with it from the knowledge of what the work produces and reproduces. Much as Tolkien's examination of ecological concerns places him in conversation with other Anthropocenic writers who came well after his time, and from whose company he has long been excluded due to both the era and the genre in which he was writing, his engagement with empire and race places him in a distinct literary context. Coming towards the tail-end of British imperialist literature, Tolkien's writing on the entanglement of power and land reveals a literary tradition on the tipping point between conservatism and a more radical politics, one that was swiftly beginning to be troubled by the legacies of empire, and understand the doubled harm of human and non-human this entails, even while it could not remove itself from the social and existential luxury such superiority afforded. The legendarium becomes, in the strange liminal way that defines so much of Tolkien's writing, and which came to define himself as a writer, a work of both anti-colonial and profoundly colonial fiction – a product of its time, both past, present and future.

Conclusion

In 'Maps, Knowledge and Power', critical cartographer J. B. Harley laments that the function of maps, as a series of images encoded with specifically historical practices, have remained 'largely undifferentiated from the wider geographical discourse in which they are often embedded', demanding instead how we might make maps '"speak" about the social worlds of the past?' (1988, 277). This book has taken critical cartography at its word, building a case for how we might make Tolkien's fictional cartographic corpus speak to the sociopolitical and cultural conditions of the worlds it depicts. In doing so, this book demands a wholesale acknowledgement of the socially constructed and value-laden imagery of the map, and its inherently political textuality. By placing literary cartography in direct conversation with the work of critical cartographers, the ways in which maps articulate and facilitate the power relations even within fictional worlds becomes apparent. The socially constructed nature of the map doubles; the diegetic map-maker decides and produces the knowledge of the map, yet Tolkien actively produces its inherent power, binding his cartographic practice tightly to the other narratives of power that wind throughout his legendarium: the power the human deploys over the non-human, the crisis of anxiety and control underlying the tension between human and non-human temporalities, the brutal human and non-human cost of imperialism and territorial conquest. Fictional maps have long been understood in relation to their accompanying text solely through their illustrative capacity; *Mapping Middle-earth* complicates this understanding, moving past a purely representational text–image relationship and revealing instead the political enmeshments that simultaneously inform both narrative and map.

This kind of knowledge production, both in the primary and secondary world, is crucial to the establishment of these power structures. Maps, as Harley argues, act as forms of Foucauldian knowledge production, presenting a means of understanding, interpreting and navigating the world that can then be used to gain control. This latent entanglement between the representational power that the map encodes and the ways that it is deployed runs throughout Tolkien's legendarium, both in the ways that he draws on historical methods of knowledge communication in maps, such as the structuring of unknown areas at the edge of the page to mimic medieval conceptualizations of spatiality as examined in the first chapter, and in the ways that maps within the legendarium designate, formalize and narrativize the land. The power relations that this enables are then made explicit. As argued in the second chapter, the knowledge that the map exemplifies acts as a manifestation of the human/nature hierarchy that has historically been predicated on derationalization and the subsequent subjugation of nature as argued by Val Plumwood, Timothy Morton, Donna Haraway and others. This hierarchy then enables various acts of power and control over the natural world that range from the reading of difficult, unwelcoming landscapes in order to successfully traverse them, as conducted by Aragorn and Bilbo, to the explicitly environmentally destructive, such as the mass industrialization that is implemented after Saruman's corruption and the deforestation of Númenor and Middle-earth during the days of the Númenórean empire. The third chapter established how maps can counteract anxieties surrounding time and temporalities, both in terms of the tension between human and non-human scales of time and the passage of natural time itself by acting as a way of artificially freezing the time and rendering it in human scales. In this case, maps also act as a means of charting and thereby rendering spatial and thus knowable other shifts taking place through time, such as the anthropological degeneration of the Elves in Middle-earth and their fading into the Undying Lands. The fourth chapter, meanwhile, concerns explicit acts of political power that are exercised over land and territory, and explicates the ways in which cartography expresses and implements these. Saïd, Fanon and Elizabeth M. DeLoughrey all illustrate the ways in which power politics and imperialism simultaneously and deliberately harm both the land and its inhabitants; the threat of political violence is present throughout Middle-earth's cartography, whether through the rewriting of borders and territorial spaces as their political allegiance shifts, or through the absences on the map such as the Drúedain, who are constantly denied and erased from their homeland.

Crucial to understanding these power relations is the tension between the map's attempt at totalizing representation and the resistance to mapping displayed by the non-human world that Tolkien encodes within his cartography. Employing the generic potentialities of the fantasy genre, Tolkien exaggerates the agency of the non-human, edging towards a posthumanist understanding of nature that begins to dissolve the boundaries between human and non-human subjectivities, in order to empower the non-human world against the power and control that cartography represents. Caradhras refuses the safe passage that the map promises by causing rock falls to impede trespassers, while the Huorns and Ents move and act with what Plumwood has theorized as 'intentionality', thus reconfiguring agency to include the actions of the non-human and refusing the passivity demanded by cartography and the control that it implies. Elsewhere, the map's attempt to freeze time and render legible non-human timescales only works to strip the maps entirely of their power and practicality; the maps in the Númenórean archive are ossified into historical material that no longer has use beyond the archive, and can only speak to the regret over the passing of time and the loss that this predicates. This resistance to the map's exercise of power does not invalidate critical cartography's emphasis on the innately political nature of maps, or Foucault's framework of inevitable and inescapable power structures; rather, these illustrations of the occasional futility of mapping not only highlight the agency that these power structures seek to quell but also act as a form of creative protest, where the generic devices of the fantasy genre enable a negation of the dominance of the human world that is otherwise not possible.

For, as much as this is a book about mapping, it is also a book about time. Cartography, as Harley argues, is a text that encodes both space and time, both in the map's temporal language and in its representation of context; maps work to speak to the social worlds of the past, present and future, revealing as much about the social, political and cultural milieu in which they were created as they do about the geography of the land they depict. In the case of Tolkien's cartographic corpus, the maps speak to the social worlds that he created; yet they also speak to his own world, and the contemporary anxieties that scaffolded it. Writing as he was through the bulk of the twentieth century, in the short years preceding the emergence of environmental literature, Anthropocene discourse and the dawning realization of our ongoing ecological crisis, Tolkien's legendarium demonstrates an explicit engagement with this burgeoning context that would come to define the decades that immediately followed the publication of his work. Similarly, his legendarium is situated firmly not only in Victorian and post-Victorian narratives of scientific progress and the discoveries of deep time, but in the concomitant dislocation of human and non-human timescales, and the anxieties of fragmentation, disempowerment and irrelevance that have been central to literature since the beginnings of Modernism. His engagement with territorial conquest, meanwhile, while lacking the same radical thrust as his environmental thought did, continues to firmly situate him as a writer of his time: one produced by a profoundly colonial context and the white supremacy of the British Empire, and one whose discomfort with imperial power made itself known through inquisitive yet imperfectly anti-colonial tales.

There are multiple threads to follow in the wake of this book. There exists, for example, a corpus of adapted maps of Tolkien's legendarium, amongst them Karen Wynn Fonstad's *The Atlas of Middle-earth*, Barbara Strachey's *Journeys of Frodo*, Pauline Baynes' 'Map of Middle-earth', and the digital visualization LotrProject, which features interactive maps of the first and third Ages of Middle-earth that can be customized to filter places, events and the journey paths of different characters. All of these might be considered as a mode of adaptation, a practice that Linda Hutcheon defines as both a product and process that prunes, morphs and narrates the original text, creating a form of knowledge production all of its own (2006, 21). There is also plenty of ground to further examine each of these chapters more broadly within the field of Tolkien studies, and the literary perspectives that might be gained by positioning Tolkien's critiques of environment, colonialism and race within contemporary theoretical discourse. In addition, there is potential for a study that takes this intersection between geocriticism and literary criticism and applies it either across a spectrum of fantasy maps or literary cartography more broadly conceived. It is true that Tolkien's intricate world-building, his large corpus of narrative and posthumously published maps and his conceit that all of his texts were diegetically produced particularly enable this kind of analysis; however, as outlined in the Introduction, there existed an extensive corpus of literary cartography even before Tolkien, and as Stefan Ekman has highlighted, almost innumerable maps since, all of which have the potential to be read through this lens. Critical cartography seeks to emphasize the innate political practice of the map; any study of fictional cartography therefore needs to recognize the ways in which, as Denis Wood argues, power constitutes the ability to do work, and the ways in which all maps – even fictional ones – work.

More than anything, however, I hope this book makes a case for the need to situate Tolkien within time; both his own contemporary context – which was profoundly more political than certain imaginations often allow – and within our own contemporary literary discourses, in order to bring his oeuvre more seriously within the field of literary studies. Reading Tolkien's legendarium, particularly in the last several years, can feel so often like a premonition; an oracle into the ways in which the fundamental cultural and political ideas that we have built around the non-human world – its irrationality, its disposability, its unfeelingness – might lead to a complex and comprehensive state of destruction.

It is read in his entanglement between ecological and imperialist violence, his obsession with moments of geological collapse, in the ongoing, tangled legacies of haunting and loss that thread through the lands and peoples of Middle-earth. It is only by placing Tolkien within our own still evolving, searching, contemporary understandings of these collective mistake-makings that we might come to see the ways in which he grasped – ahead of his time – so much of it. For a text set so entirely, almost more than any other before or since, in a wholly distinct and fully realized alternative space and time, it is remarkable, and deeply necessary to understand, how much it is also about our own.

Bibliography

Bennett, Jane. 2010. *Vibrant Matter: A Political Ecology of Things*. Durham, NC: Duke University Press.

Berghoff, Hartmut, and Barbara Korte. 2002. 'Britain and the Making of Modern Tourism: An Interdisciplinary Approach'. In *The Making of Modern Tourism: The Cultural History of the British Experience, 1600–2000*, edited by Hartmut Berghoff, Barbara Korte, Ralf Schneider, and Christopher Harvie, 1–20. Basingstoke: Palgrave Macmillan.

Black, Jeremy. 1997. *Maps and Politics*. Chicago: University of Chicago Press.

Black, Jeremy. 2016. 'War and Cartography'. In *Maps and the 20th Century: Drawing the Line*, edited by Tom Harper, 31–66. London: The British Library.

Bodleian Library. 1992. *J.R.R. Tolkien: Life and Legend: An Exhibition to Commemorate the Centenary of the Birth of J.R.R. Tolkien (1892–1973)*. Oxford: Bodleian Library.

Borges, Jorge Luis. 2018. 'On Exactitude in Science'. In *The Garden of Forking Paths*, 35. London: Penguin.

Bowler, Peter J. 2003. *Evolution: The History of an Idea*. Berkeley, CA: University of California Press.

Braidotti, Rosi. 2013. *The Posthuman*. Oxford: Polity Press. http://ebookcentral.proquest.com/lib/ed/detail.action?docID=1315633.

Breen, Katharine. 2005. 'Returning Home from Jerusalem: Matthew Paris's First Map of Britain in Its Manuscript Context'. *Representations; Berkeley*, no. 89 (Winter): 59–93.

Bubandt, Nils. 2017. 'Haunted Geologies: Spirits, Stones, and the Necropolitics of the Anthropocene'. In *Arts of Living on a Damaged Planet*, edited by Anna Lowenhaupt Tsing, Heather Swanson, Elaine Gan, and Nils Bubandt, 121. Minneapolis: University of Minnesota Press.

Bush, Barbara. 2014. *Imperialism and Postcolonialism*. Abingdon: Routledge.

Cadbury, Deborah. 2000. *The Dinosaur Hunters: A True Story of Scientific Rivalry and the Discovery of the Prehistoric World*. London: HarperCollins UK.

Camille, Michael. 1992. *Image on the Edge: The Margins of Medieval Art*. Essays in Art and Culture. London: Reaktion Books.

Campbell, Tony. 1987. 'Portolan Charts from the Late Thirteenth Century to 1500'. In *The History of Cartography*, Vol. 1: *Cartography in Prehistoric, Ancient, and Medieval Europe and the Mediterranean*, edited by J. B. Harley and David Woodward, 371–463. Chicago: University of Chicago Press.

Carpenter, Humphrey. 2000. *J.R.R. Tolkien: A Biography*. New York: Houghton Mifflin (first published 1977).

Carroll, Lewis. 1982. *The Penguin Complete Lewis Carroll*. Harmondsworth: Penguin.

Carson, Rachel. 1962. *Silent Spring*. Boston, MA: Houghton Mifflin Company.
Cecire, Maria Sachiko. 2016. 'English Exploration and Textual Travel in The Voyage of the Dawn Treader'. In *Space and Place in Children's Literature, 1789 to the Present*, edited by Maria Sachiko Cecire, Hannah Field, Kavita Mudan Finn, and Malini Roy, 111–28. Abingdon: Routledge.
Chasseaud, Peter. 2013. *Mapping the First World War*. Glasgow: Collins.
Clarke, Bruce, and Manuela Rossini. 2017. 'Preface: Literature, Posthumanism and the Posthuman'. In *The Cambridge Companion to Literature and the Posthuman*, edited by Bruce Clarke and Manuela Rossini, xi–xxii. Cambridge: Cambridge University Press.
Clifford, Anne M. 1991. 'Creation'. In *Systematic Theology: Roman Catholic Perspectives*, edited by Francis Schüssler Fiorenza and John P. Galvin, 193–248. Minneapolis, MN: Fortress Press.
Cohen, Jeffrey Jerome. 2015. *Stone: An Ecology of the Inhuman*. Minneapolis: University of Minnesota Press.
Connolly, Daniel K. 2009. *The Maps of Matthew Paris: Medieval Journeys through Space, Time and Liturgy*. Woodbridge: The Boydell Press.
Conrad, Joseph. 1971. *Heart of Darkness: An Authoritative Text, Backgrounds and Sources, Criticism*. Edited by Robert Kimbrough. Revised edition. Norton Critical Edition. New York: Norton.
Cooper, David. 2012. 'Critical Literary Cartography: Text, Maps and a Coleridge Notebook'. In *Mapping Cultures: Place, Practice, Performance*, edited by Les Roberts, 29–52. London: Palgrave Macmillan UK.
Cosgrove, Denis. 2007. 'Mapping the World'. In *Maps: Finding Our Place in the World*, edited by James R. Akerman and Robert W. Karrow Jr, 65–116. Chicago: University of Chicago Press.
Croft, Janet Brennan, and Leslie A. Donovan, editors. 2015. *Perilous and Fair: Women in the Works and Life of J.R.R. Tolkien*. Altadena, CA: Mythopoeic Press, 2015.
Cuddy-Keane, Melba. 2002. 'Imaging/Imagining Globalization: Maps and Models'. In *Society for Critical Exchange*. New York. http://homes.chass.utoronto.ca/-mcuddy/mapping.htm.
Daniels, Stephen, Dydia Delyser, J. Nicholas Entrikin, and Douglas Richardson. 2011. 'Introduction'. In *Envisioning Landscapes, Making Worlds: Geography and the Humanities*, edited by Stephen Daniels, Dydia Delyser, J. Nicholas Entrikin, and Douglas Richardson, xxvi–xxxii. Hoboken, NJ: Taylor & Francis.
De Quincey, Thomas. 1967. *The Confessions of an English Opium-Eater*. London: Everyman's Library.
Delano Smith, Catherine. 1987. 'Cartography in the Prehistoric Period in the Old World: Europe, the Middle East, and North Africa'. In *The History of Cartography, Vol. 1: Cartography in Prehistoric, Ancient, and Medieval Europe and the Mediterranean*, edited by J. B. Harley and David Woodward, 54–101. Chicago: University of Chicago Press.

Delano Smith, Catherine, and Roger J. P. Kain. 1999. *English Maps: A History*, vol. 2. The British Library Studies in Map History. London: The British Library.

DeLoughrey, Elizabeth M., and George B. Handley. 2011. 'Introduction: Towards an Aesthetics of the Earth'. In *Postcolonial Ecologies Literatures of the Environment*, edited by Elizabeth M. DeLoughrey and George B. Handley, 3–39. Oxford: Oxford University Press.

Dickerson, Matthew, and Jonathan Evans. 2006. *Ents, Elves, and Eriador: The Environmental Vision of J.R.R. Tolkien*. Lexington: University Press of Kentucky.

Doyle, Arthur Conan. 1974. *The Case-Book of Sherlock Holmes*. London: John Murray Publishers.

Doyle, Arthur Conan. 1980. *Sherlock Holmes: Selected Stories*. Oxford: Oxford University Press.

Eco, Umberto. 1986. *Travels in Hyperreality: Essays*. Picador. New York: Harcourt.

Edney, Matthew H. 1997. *Mapping an Empire: The Geographical Construction of British India, 1765–1843*. Chicago: University of Chicago Press.

Ekman, Stefan. 2013. *Here Be Dragons: Exploring Fantasy Maps and Settings*. Middletown, CT: Wesleyan University Press.

Espenhorst, Jürgen. 2016. 'A Good Map Is Half the Battle! The Military Cartography of the Central Powers in World War I'. In *History of Military Cartography*, edited by Elri Liebenberg, Imre Josef Demhardt, Soetkin Vervust, 83–130. Cham: Springer.

Fanon, Frantz. 1963. *The Wretched of the Earth*. Translated by Constance Farrington. New York: Grove Press.

Farrier, David. 2019. *Anthropocene Poetics: Deep Time, Sacrifice Zones, and Extinction*. Posthumanities 50. Minneapolis: University of Minnesota Press.

Fimi, Dimitra. 2009. *Tolkien, Race and Cultural History: From Fairies to Hobbits*. London: Palgrave Macmillan.

Fisher, Jason. 2010. 'Sourcing Tolkien's "Circles of the World": Speculations on the Heimskringla, the Latin Vulgate Bible, and the Hereford Mappa Mundi'. In *Middle-Earth and Beyond: Essays on the World of J.R.R. Tolkien*, edited by Janka Kaščáková, 1–18. Newcastle-upon-Tyne: Cambridge Scholars Publishing.

Flieger, Verlyn. 2000. 'Taking the Part of Trees: Eco-Conflict in Middle-Earth'. In *J.R.R. Tolkien and His Literary Resonances*, edited by George Clark and Daniel Timmons, 145–58. Westport, CT: Greenwood.

Flieger, Verlyn. 2001. *A Question of Time: J.R.R. Tolkien's Road to Faërie*. Kent: Kent State University Press.

Flieger, Verlyn. 2003. 'Tolkien's Wild Men: From Medieval to Modern'. In *Tolkien the Medievalist*, edited by Jane Chance, 95–105. New York: Routledge.

Flieger, Verlyn. 2013. 'The Forests and the Trees: Sal and Ian in Faërie'. In *Tolkien: The Forest and the City*, edited by Gerard Hynes and Helen Conrad O'Briain, 107–22. Dublin, Ireland: Four Courts Press.

Flynn, Thomas. 2016. 'Foucault among the Geographers'. In *Space, Knowledge and Power: Foucault and Geography*, edited by Jeremy W. Crampton and Stuart Elden, 59–66. Oxford: Routledge.

Fonstad, Karen Wynn. 1994. *The Atlas of Middle-Earth*. London: HarperCollins Publishers.
Foucault, Michel. 1978. *The History of Sexuality*, vol. 1: *An Introduction*. Translated by Robert Hurley. New York: Pantheon Books.
Foucault, Michel. 1995. *Discipline and Punish: The Birth of the Prison*. London: Vintage Books.
Foucault, Michel. 2002. *The Archaeology of Knowledge*. Translated by A. M. Sheridan Smith. Oxford: Routledge.
Foucault, Michel. 2007. 'The Meshes of Power'. In *Space, Knowledge and Power: Foucault and Geography*, edited by Jeremy W. Crampton and Stuart Elden, translated by Gerald Moore, 153–62. Aldershot: Ashgate.
Garrard, Greg. 2004. *Ecocriticism*. London: Routledge.
Genette, Gerard. 1997. *Paratexts: Thresholds of Interpretation*. Translated by Jane E. Lewin. Cambridge: Cambridge University Press.
Glissant, Édouard. 1997. *The Poetics of Relation*. Ann Arbor: The University of Michigan Press.
Gould, Stephen Jay. 1987. *Time's Arrow, Time's Cycle: Myth and Metaphor in the Discovery of Geological Time*. Cambridge, MA: Harvard University Press.
Haggard, H. Rider. 1998. *King Solomon's Mines*. Oxford: Oxford University Press.
Hammond, Wayne G., and Christina Scull. 1995. *J.R.R. Tolkien: Artist and Illustrator*. London: HarperCollins Publishers.
Hammond, Wayne G., and Christina Scull. 2006. *The J.R.R. Tolkien Companion & Guide: Chronology*. London: HarperCollins Publishers.
Hammond, Wayne G., and Christina Scull. 2011. *The Art of the Hobbit by J.R.R. Tolkien*. London: HarperCollins Publishers.
Hammond, Wayne G., and Christina Scull. 2015. *The Art of the Lord of the Rings by J.R.R. Tolkien*. London: HarperCollins Publishers.
Haraway, Donna J. 2016. *Staying with the Trouble*. Durham, NC: Duke University Press.
Harley, J. B. 1988. 'Maps, Knowledge, and Power'. In *The Iconography of Landscape: Essays on the Symbolic Representation, Design and Use of Past Environments*, edited by Denis Cosgrove and Stephen Daniels, 277–312. Cambridge: Cambridge University Press.
Harley, J. B. 1990. 'Text and Contexts in the Interpretation of Early Maps'. In *From Sea Charts to Satellite Images: Interpreting Noth American History through Maps*, edited by David Buisseret, 3–15. Chicago: University of Chicago Press.
Harley, J. B. 2001. 'Deconstructing the Map'. In *The New Nature of Maps*, edited by Paul Laxton, 149–68. Baltimore, MD: Johns Hopkins University Press.
Harrison, Robert Pogue. 1992. *Forests: The Shadow of Civilization*. Chicago: University of Chicago Press.
Harvey, P. D. A. 1987a. 'Local and Regional Cartography in Medieval Europe'. In *The History of Cartography*, vol. 1: *Cartography in Prehistoric, Ancient, and Medieval Europe and the Mediterranean*, edited by J. B. Harley and David Woodward, 464–501. Chicago: University of Chicago Press.

Harvey, P. D. A. 1987b. 'Medieval Maps: An Introduction'. In *The History of Cartography*, vol. 1: *Cartography in Prehistoric, Ancient, and Medieval Europe and the Mediterranean*, edited by J. B. Harley and David Woodward, 283–5. Chicago: University of Chicago Press.

Harvey, P. D. A. 1991. *Medieval Maps*. London: The British Library.

Heffernan, Michael. 1996. 'Geography, Cartography and Military Intelligence: The Royal Geographical Society and the First World War'. *Transactions of the Institute of British Geographers* 21 (3): 504–33.

Heringman, Noah. 2015. 'Deep Time at the Dawn of the Anthropocene'. *Representations* 129 (1): 56–85.

Hewitt, Rachel. 2010. *Map of a Nation: A Biography of the Ordnance Survey*. London: Granta Publications.

Hoiem, Elizabeth Massa. 2005. 'World Creation as Colonization: British Imperialism in "Aldarion and Erendis"'. *Tolkien Studies* 2 (1): 75–92.

Hunt, Peter. 1987. 'Landscapes and Journeys, Metaphors and Maps: The Distinctive Feature of English Fantasy'. *Children's Literature Association Quarterly* 12 (1): 11–14.

Hutcheon, Linda. 2006. *A Theory of Adaptation*. Abingdon: Routledge.

Hynes, Gerard. 2012. '"Beneath the Earth's Dark Keel": Tolkien and Geology'. *Tolkien Studies* 9 (1): 21–36.

Hynes, Gerard. 2013. '"The Cedar Is Fallen": Empire, Deforestation and the Fall of Númenor'. In *Tolkien: The Forest and the City*, edited by Helen Conrad O'Briain and Gerard Hynes, 123–32. Dublin, Ireland: Four Courts Press.

Jeffers, Susan. 2014. *Arda Inhabited: Environmental Relationships in The Lord of the Rings*. Kent, OH: Kent State University Press.

Jones, Diana Wynne. 2004. *The Tough Guide to Fantasyland*. London: Gollancz (first published by Vista Books (London) in 1996).

Kitchin, Rob, Chris Perkins and Martin Dodge. 2009. 'Thinking about Maps'. In *Rethinking Maps: New Frontiers in Cartographic Theory*, edited by Martin Dodge, Rob Kitchin and Chris Perkins, 1–25. London: Routledge.

Lewis, Suzanne. 1987. *The Art of Matthew Paris in the Chronica Majora*. Aldershot: Gower in association with Corpus Christi College, Cambridge.

Lewis-Jones, Huw. 2018. *The Writer's Map: An Atlas of Imaginary Lands*. London: Thames and Hudson.

Lilley, Keith D. 2011. 'Digital Cartographies and Medieval Geographies'. In *Envisioning Landscapes, Making Worlds: Geography and the Humanities*, edited by Stephen Daniels, Dydia Delyser, J. Nicholas Entrikin and Douglas Richardson, 25–33. Hoboken, NJ: Taylor & Francis.

Macfarlane, Robert. 2016. *Landmarks*. London: Penguin Books.

Mendlesohn, Farah. 2008. *Rhetorics of Fantasy*. Middletown, CT: Wesleyan University Press.

Mitchell, J. B. 1933. 'The Matthew Paris Maps'. *The Geographical Journal* 81 (1): 27–34.

Morton, Timothy. 2007. *Ecology without Nature: Rethinking Environmental Aesthetics*. Cambridge, MA: Harvard University Press.

Nixon, Rob. 2005. 'Environmentalism and Postcolonialism'. In *Postcolonial Studies and Beyond*, edited by Ania Loomba, Suvir Kaul, Matti Bunzl, Antoinette Burton and Jed Esty, 233–51. Durham, NC: Duke University Press.

Obertino, James. 2006. 'Barbarians and Imperialism in Tacitus and The Lord of the Rings'. *Tolkien Studies* 3 (1): 117–31.

Oliver, Richard. 2005. *Ordnance Survey Maps: A Concise Guide for Historians*. London: The Charles Close Society.

Padrón, Ricardo. 2007. 'Mapping Imaginary Worlds'. In *Maps: Finding Our Place in the World*, edited by James R. Akerman and Robert W. Karrow Jr, 255–88. Chicago, IL: The University of Chicago Press.

Parrlla, Sofia. 2021. 'All Worthy Things: The Personhood of Nature In J.R.R. Tolkien's Legendarium'. *Mythlore* 40 (1): 5–20.

Plumwood, Val. 1993. *Feminism and the Mastery of Nature*. Opening Out. London: Routledge.

Poster, Mark. 1982. 'Foucault and History'. *Social Research* 49 (1): 116–42.

Rateliff, John D. 2006. '"And All the Days of Her Life Are Forgotten": The Lord of the Rings as Mythic Prehistory'. In *The Lord of the Rings 1954-2004: Scholarship in Honor of Richard E. Blackwelder*, edited by Wayne G. Hammond and Christina Scull, 67–100. Milwaukee, WI: Marquette University Press.

Rateliff, John D. 2011. *The History of the Hobbit*. Revised edition. London: HarperCollins.

Riffenburgh, Beau. 2014. *Mapping the World: The Story of Cartography*. London: Carlton Publishing Group.

Rossetto, Tania. 2014. 'Theorizing Maps with Literature'. *Progress in Human Geography* 38 (4): 513–30.

Saïd, Edward W. 1990. 'Yeats and Decolonization'. In *Nationalism, Colonialism, and Literature*, edited by Terry Eagleton and Fredric Jameson, 69–96. Minneapolis: University of Minnesota Press.

Saïd, Edward W. 1994. *Culture and Imperialism*. London: Vintage.

Saïd, Edward W. 2003. *Orientalism*. London: Penguin Books.

Sessions, George. 1995. 'Preface'. In *Deep Ecology for the Twenty-First Century*, edited by George Sessions, ix–xxviii. Boston, MA: Shambhala.

Sharpe, Christina Elizabeth. 2016. *In the Wake: On Blackness and Being*. Durham, NC: Duke University Press. https://doi.org/10.1515/9780822373452.

Shippey, Tom. 2005. *The Road to Middle-Earth*. 3rd edition. London: HarperCollins Publishers.

Stevenson, Robert Louis. 1895. 'My First Book'. Edited by Jerome Klapka Jerome. *The Idler; an Illustrated Magazine* 6: 3–11. London: Chatto & Windus.

Tally, Robert T. 2013. *Spatiality*. New Critical Idiom. London: Routledge.

Thacker, Andrew. 2005. 'The Idea of a Critical Literary Geography'. *New Formations* 57 (1): 56–73.
Tolkien, J. R. R. 2018. *The Hobbit Facsimile Gift Edition*. First Thus edition. London: HarperCollins.
Tolkien, J. R. R. 1983. *The Book of Lost Tales: Part One*. London: HarperCollins Publishers.
Tolkien, J. R. R. 1986. *The Shaping of Middle-Earth*. London: HarperCollins Publishers.
Tolkien, J. R. R. 1987. *The Lost Road and Other Writings*. London: HarperCollins Publishers.
Tolkien, J. R. R. 1989. *The Treason of Isengard*. London: HarperCollins Publishers.
Tolkien, J. R. R. 1990. *The War of the Ring*. London: HarperCollins Publishers.
Tolkien, J. R. R. 1991. *Unfinished Tales*. London: Grafton.
Tolkien, J. R. R. 1992. *Sauron Defeated*. London: HarperCollins Publishers.
Tolkien, J. R. R. 1993. *Morgoth's Ring*. London: HarperCollins Publishers.
Tolkien, J. R. R. 1994. *The War of the Jewels*. London: HarperCollins Publishers.
Tolkien, J. R. R. 1996. *The Peoples of Middle-Earth*. London: HarperCollins Publishers.
Tolkien, J. R. R. 1997a. 'A Secret Vice'. In *The Monsters and the Critics: And Other Essays*, edited by Christopher Tolkien, 198–223. London: HarperCollins Publishers.
Tolkien, J. R. R. 1997b. *The Monsters and the Critics: And Other Essays*. London: HarperCollins UK.
Tolkien, J. R. R. 2003. *The Annotated Hobbit: The Hobbit, or, There and Back Again*. London: HarperCollins Publishers.
Tolkien, J. R. R. 2006. *The Letters of J.R.R. Tolkien*. Edited by Humphrey Carpenter. London: HarperCollins Publishers.
Tolkien, J. R. R. 2008a. *The Fellowship of the Ring*. London: HarperCollins Publishers.
Tolkien, J. R. R. 2008b. *The Hobbit*. London: HarperCollins Publishers.
Tolkien, J. R. R. 2008c. *The Return of the King*. London: HarperCollins Publishers.
Tolkien, J. R. R. 2008d. *The Silmarillion*. London: HarperCollins.
Tolkien, J. R. R. 2008e. *The Two Towers*. London: HarperCollins Publishers.
Tolkien, J. R. R. 2021. *The Nature of Middle-earth*. Edited by Carl F. Hostetter. London: HarperCollins Publishers and Mariner Books.
Tsing, Anna Lowenhaupt, Heather Swanson, Elaine Gan, and Nils Bubandt. 2017. 'Introduction'. In *Arts of Living on a Damaged Planet*, edited by Anna Lowenhaupt Tsing, Heather Swanson, Elaine Gan and Nils Bubandt, 1–16. Minneapolis: University of Minnesota Press.
Turchi, Peter. 2004. *Maps of the Imagination: The Writer as Cartographer*. San Antonio, TX: Trinity University Press.
Vaninskaya, Anna. 2020. *Fantasies of Time and Death: Dunsany, Eddison, Tolkien*. London: Palgrave Macmillan.
Walker, R. C. 1981. 'The Cartography of Fantasy'. *Mythlore* 7 (4): 37–8.
Williams, John. 1997. 'Isidore, Orosius and the Beatus Map'. *Imago Mundi* 49: 7–32.
Wood, Denis. 1992. *The Power of Maps*. London: Routledge.

Wood, Denis, and John Fels. 2008. *The Natures of Maps*. Chicago, IL: University of Chicago Press.

Woodward, David. 1987. 'Medieval Mappaemundi'. In *The History of Cartography*, vol. 1: *Cartography in Prehistoric, Ancient, and Medieval Europe and the Mediterranean*, edited by J. B. Harley and David Woodward, 286–370. Chicago, IL: University of Chicago Press.

Yusoff, Kathryn. 2018. *A Billion Black Anthropocenes or None*. Minneapolis: University of Minnesota Press.

Index

Aldarion 85–88, 156, 163, 167–9
Ambarkanta 24, 46–9, 105–6, 108, 112–13, 129–30
    diagrams and maps (see Tolkien, J. R. R. Cartography)
Anthropocene 19, 65–8, 73, 97, 99, 101–2, 104, 109–10, 112, 133–4, 139, 173
anthropocentrism 66–9, 73, 78, 80, 84, 89, 92–4, 102, 108, 122, 135, 167
anthropomorphization 92–4
anti-colonial 136, 156, 160, 170, 174
Atlantis 107–8, 110

Barthes, Roland 10
Baynes, Pauline 14, 25, 174
Beleriand
    destruction of 106–9, 113, 151
    map (*see* Tolkien, J. R. R. Cartography)
Beowulf 118
borders 17, 42, 74, 79, 88, 95–6, 122, 135, 140, 143–150, 154, 162–3, 169, 172
Borges, Jorge Luis 21–2, 32, 35–6, 44
British Empire 3, 136, 140, 155, 170, 174

Caradhras 89, 91–4, 96, 112, 173
Carroll, Lewis 13, 21–2, 32, 35, 36
Carson, Rachel 66
cartography
    accuracy 2, 9, 21–2, 28–31, 35–8, 40–3, 48–9, 54, 57–61, 73, 102, 125, 154, 158n
    blank spaces 30, 39, 161–2
    representation 8–11, 15–16, 19, 21–23, 26, 27, 29–41, 43, 47–49, 57–60, 65, 72, 75, 90, 98, 102, 112, 114, 117–18, 120, 125, 129–30, 139–40, 147, 154, 157, 171–3
    critical 3, 5, 8, 11–2, 18–19, 66, 70n, 102, 139–40, 171, 173–4
    edges of 30–1, 52–3, 162
    military 24, 37–38, 40–3, 57, 60–1
    navigational 31–4, 38, 40, 51, 56, 114, 146, 149
    palimpsest 43, 162
    religious 28–30, 43
    temporal 19, 34, 102, 112–4, 118, 120, 125, 130–1, 154, 172
catastrophism 98–100, 102, 104–9, 114
chthonic 68–9, 72, 91, 94–5
Chthulucene 68
colonialism 5, 11, 17–18, 134–8, 155–6, 160–1, 163, 165–70, 174
Conrad, Joseph 162, 166
contour lines 39–40, 42, 47, 55–6, 58–60
Cuvier, Georges 115, 117

deep ecology 66–67, 69, 71
deep time 3, 5, 19, 61, 97–109, 112, 114, 115, 117–18, 120, 122–5, 131, 173
deforestation 86–8, 97, 138, 166–9, 172
displacement 136, 138–9, 157–61
Doyle, Arthur Conan 14, 159
Drúadan Forest 60, 158n, 160
Drúedain 150, 156–61, 165, 169, 172

Ebstorf map 28–30, 45, 48, 52–3
ecological crisis 5, 19, 64, 68, 97, 110, 133, 169, 173
ecological damage 3, 65, 72, 80, 82–9, 97, 99, 105, 108–10, 133–136, 138, 150–3, 166–7, 169–70, 172
Elder Days 103, 122–3
Enlightenment 35–9, 139
Ents 83, 93–6, 104
environmental racism 19, 138, 166, 169
Erendis 85–87, 156, 163, 168
evolution 19, 100n

fading of the Elves 124–30, 172
fellbeasts 116
fossils 101, 104, 114–18, 120, 126
Foucault, Michel 5–10, 17–18, 70n, 140, 146, 171, 173

geographical violence 137, 138, 140, 152, 154, 160-1
geology 46, 48-9, 61, 65-6, 97-119, 126, 129, 131, 133-5, 139, 175
Ghân-Buri-Ghân 158-60
Girdle of Melian 148
Golden Spike 65, 139
Gondolin 58, 145-7
Gough Map of Britain 33-4, 51, 58

hachures 39, 49-51, 54-5
Haggard, H. Rider 14, 159
Haraway, Donna 19, 67-8, 91-2, 95, 172
Hereford map 24, 28-30, 48, 52
Huorns 83, 93-96, 173
Hutton, James 19, 99-101, 103

immortality 104, 107, 109, 119-30, 167, 169
imperialism 3, 5, 19, 58, 61, 133-40, 149, 151-2, 155-170, 171-2, 174-5
indigenous peoples 134, 138-9, 160, 162
industrialization 3, 38, 63, 65, 74, 82, 131, 134, 152-3, 172
Industrial Revolution 65, 134, 139
Isengard 83, 88, 94-6, 163
Isidorian maps 27, 28, 48
itinerary map 33-4, 51-2, 58

Jones, Diana Wynne 12, 15

Lewis, C. S. 14, 16, 25, 93
literacy of land 76-7, 101
Lond Daer 87-8, 166, 168
Lonely Mountain, the 4, 24, 49, 51-2, 73-4, 76, 135, 146-51, 161
Lothlórien 122-5, 154
Lyell, Charles 19, 99-100

Macrobius maps 27-28, 48
Maginot Line 145
Matthew Paris' maps of Britain 33-4, 51
Mercator projection 10
Mirkwood 50-3, 55, 58, 73-6, 89-91, 95
Mappae mundi 24, 27-32, 34, 46, 48, 52-3, 58

Narnia books 14, 17, 25, 93
necropolitics 110-11
Non-human

agency 19, 68, 88-96, 104, 107-108, 112, 118, 172-3
binary with the human 18-9, 37, 64, 66-7, 69, 71-7, 91, 96, 134
instrumentalization 66, 69, 71-2, 76, 81, 95, 141, 168
nuclear weapons 65, 109, 139
Númenor
  drowning of, or Akallabêth 47, 106-14, 119-21, 129, 148, 169
  map (*see* Tolkien, J. R. R. Cartography)

Old Forest, the 17, 55, 71-2, 82-4, 89-91, 95
Old Man Willow 71, 82-3, 89
Ordnance Survey 25, 37-41, 43, 57-8
Orosian maps 27-8

Panopticon 7
paratext 2-4, 15, 17-18, 44, 46, 51, 60-1, 113, 140, 168
pastoral 63, 71, 74, 82-3
Pelennor Fields 79, 116, 144-5, 160
Playfair, James 101, 122
portolan charts 31-2, 35, 48
postcolonialism 4, 19, 70n, 135-9, 160
postcolonial ecocriticism 19, 138-9
posthumanism 18, 64-8, 91, 95, 172
prehistory 26, 100n, 115-7, 128
Prose Edda 45, 105
Psalter map 28-30
pseudomedievalism 25, 49, 54, 58, 148

Rammas Echor 144-5, 147
rationality 36, 65-7, 70, 72-5, 77, 91, 93, 96, 102, 123, 165
The Red Book of Westmarch 4-5

The Scouring of the Shire 82, 119, 151-5
Siccar Point 101, 122
slave trade 65, 139
Smaug 50-2, 148-51, 161
Stevenson, Robert Louis 1-2, 14
stewardship 78-88
stone-giants 93
The Straight Path 47, 129-30

Tal-Elmar 163-6, 168-9
tectonic 105-6, 11

Tolkien, Christopher 1–2, 44–46n, 53–6,
58–60, 111, 113, 121, 127, 155,
158n
Tolkien, J.R.R
   cartography
      A Part of the Shire 4, 17, 55, 57, 89,
90, 95, 154
      Ambarkanta diagrams and
maps 24, 46–9, 112–13, 129–30
      I Vene Kemen map 23, 44–6, 48–9,
53, 129
      map of Beleriand 44, 112–13, 146,
148, 168
      map of Middle-earth 53–8, 60,
88–9, 112–3, 125, 146–7, 157,
162, 168
      map of Rohan, Gondor and Mordor
53, 55, 59–61, 146–7, 158n
      map of Númenor 44, 112, 118, 168
      The West of Middle-earth at the
End of the Third Age 56, 88,
113, 118
      Thror's Map 4, 24, 49–53, 56, 58, 73,
75, 90, 114, 146–7, 149
      map of the Wilderland 49–50, 56,
90, 91, 95, 149–50
      drafts of maps 50, 52–6, 59, 150
   Catholicism 83, 100n
   relationship with nature 25, 63, 68, 78,
115, 131
   relationship with war 24, 43, 155
T-O maps 27, 29, 48
Tom Bombadil 68–72, 76, 82, 83, 104
Tookland 143
Two Trees, the 23, 45, 84–5, 87

Ungoliant 85
uniformitarianism 19, 98–102, 105–9

Valar 45, 47, 84, 103–8, 120, 135
Valinor 45, 47, 84–5, 87, 103, 107–8,
120–1, 124, 126–7, 129–30, 135,
150, 167

The White Tree 80–1
World War I 16, 24, 40–3, 60, 120,
145
World War II 120, 155